T0269499

Property rights after market liberalization reforms: land titling and investments in rural Peru

Property rights after market liberalization reforms: land titling and investments in rural Peru

Ricardo Fort

ISBN 978-90-8686-057-9

First published, 2007

Wageningen Academic Publishers
The Netherlands, 2007

Painting cover:
'Que en mi pais la gente sea feliz aunque no tenga permiso' by Pancho Guerra-Garcia

Table of contents

Acknowledgements

This thesis would not have been possible without the help and constant support of many people to whom I want to express my gratitude. I was fortunate to learn my first research steps under the supervision of many great bosses at GRADE. My sincere gratitude to Eduardo Zegarra, Alberto Pasco-Font, Manuel Glave, and Javier Escobal for sharing your knowledge with me. Many thanks also to Michael Carter for his great support during my stay at Wisconsin University.

My first meeting with Ruerd Ruben in 2003 marked the start of this long academic journey. I am enormously thankful to him for giving me his trust, and for his constant teaching and supervision. His ability to know exactly when to be flexible, and when to push harder to obtain results, has been fundamental for getting me always on track. My gratitude goes also to Arie Kuyvenhoven who has been always there to listen and advice. His guidance on the structure and implications of this study, even under difficult circumstances, has been invaluable. The motivation provided by the constant research curiosity of Javier Escobal was also of great help for completing this work. Thanks Javier for accepting the challenge.

Working at the Development Economics Group of Wageningen University has been a pleasant and unforgettable experience. I would like to thank all my colleagues, professors, and the staff for creating such a comfortable working environment. Despite its small size, Wageningen gave me many big friends. Special gratitude goes to Aureliano, Fernando, Lucia, and Elena who where there in the ups and downs. Guillermo Zuñiga has been a challenging colleague from the beginning and an excellent friend during all times; *meneco*, thanks for the team-work.

Part of my thesis work took place at the Center for Latin American Research and Documentation-CEDLA in Amsterdam. I would like to thank Annelies Zoomers and all my colleagues there for their interest in my academic work. I am also grateful to my friends Lorena, Julian, Mariana, Griet, Roxana and Carla, who made me feel at home.

The Center for International Development Issues-CIDIN of Radboud University Nijmegen welcomed me on the last year of my work in The Netherlands. I would like to thank all researchers and staff of the center for their academic and logistic support during this time. My special appreciation goes to my officemate Marloes Verhoeven for coping with me under the stress of the final months, and still keeping a smile.

To my Peruvian closest friends Tato, Pancho, Alvaro, Ernesto, Quique, Aleja, Ximena, Alexandra, and Vero who have been always there for me. Your support makes everything easier.

Finally, my deepest thanks go to my family. To my mother for her unconditional love and support, and to my sisters for always showing me the way with their example. *Papote*, I would have liked to see your proud smile more than anything. This one goes for you.

Abstract

This study discusses the links between land access, property rights, and economic development, analyzing the results and limitations of a public intervention- Land Titling and Registration- that constitutes one of the main instruments for contemporary land policy in Peru. It starts with a global perspective, and then develops a meso (or regional) and micro level approach for the study of the Peruvian Land Tilting and Registration Program (PETT). The study attempts to provide a comprehensive analysis and discussion of the importance of institutions, like land property rights, in the context of market liberalization reforms. In operational terms, this means verifying whether land titling constitutes a necessary and/or sufficient condition to promote investments and increase land values. To accomplish this objective, we use information at two different levels. We assembled a country-level panel dataset for the macro perspective, and rely on household's surveys collected during the year 2004 as part of the evaluation of the PETT Program for the micro approach of this study. Our findings reveal that titling and registration can be considered as a necessary condition to improve investment opportunities when its implementation procedure is based on the recognition of previous informal land rights and community networks, because its effect on the reduction of transaction costs at a regional level improves the dynamics of land markets and facilitates the entrance of formal financial institutions. A decentralized program is more likely to understand and correctly assess local conditions, as well as to concentrate its work on poorer farmers confronting stronger limitations to acquire tenure security by other means. Targeting must be applied also at the regional level, identifying less-developed areas that can benefit from the externality effects provided by increased levels of titling density. However, the presence of other limitations that constrain the participation of small farmers in the formal credit market, and the inability of titling to solve them by itself, makes it difficult to consider this policy as a sufficient condition to improve the livelihood of poorer farmers.

1. Introduction

1.1 Background

As many Latin-American countries, Peru experimented in the late 1960's a radical State-Led Agrarian Land Reform with the main purpose of breaking the land concentration pattern that prevailed since colonial times. Expropriated land from extensive *latifundios* was turned into production cooperatives and given to former agricultural workers to be managed under strict state supervision and planning. This production mode and its articulation with the rest of the economy collapsed after almost two decades of subsidized functioning, resulting in a strong and extremely disorganized land fragmentation process (Caballero, 1980; Matos Mar, 1980). Consequently, the new ownership pattern of agricultural land at the beginning of the 1990's was characterized by the predominance of very small landholdings and the lack of formal and clear documents of ownership over the land.

These two problems have been at the heart of the rural development policy discussion in Peru, since there is an emerging consensus about the need for enhancing the dynamics of rural land markets in a way that allows small and poor farmers to increase their productivity, improve their livelihoods and overcome poverty. The agrarian sector in Peru accounts for no more than 9 percent of the national GDP, but employs around one third of the country's working population (Instituto Nacional de Estadistica e Informatica-INEI). Moreover, according to the results of recent household surveys more than 76 percent of those living in rural areas can be considered poor under international consumption standards (Encuesta Nacional de Hogares-ENAHO, 2002).

When liberal reforms took place in Peru at the beginning of the 1990s, the percentage of parcels with a formal and registered ownership document was estimated to be less than 10 percent (Zegarra, 1999). Accordingly, starting in 1992 the Peruvian government launched a national Land Titling and Registration Program (PETT), in order to promote the formalization of property rights and improve the situation of many farmers with different types of informal documents supporting their land ownership status.

The recognition that a vast amount of the population in less-developed counties lacks formal proof of ownership for the assets they hold, and that the rules and procedures to acquire or formalize ownership and transfer rights over assets were extremely complicated, time-consuming, expensive, and hence inaccessible for the poorest segments of the population (Noronha, 1985; Platteau, 1992; Roth, 1993), prompted an immense effort by many Latin American governments and International Organizations to reform the institutions of property rights and registration systems (Deininger and Binswanger, 1999).

The renewed academic and policy attention on these matters closely reflects the reality of post-liberalization economies in Latin America where inequality and poverty persist, particularly in the rural sector (Birdsall and Sabot, 1998). The classical 'agrarian question'[1] regarding the evolution of land access for the rural poor remains today a fundamental issue for these societies, not only in economic terms but also as a social stability concern.

However, notwithstanding the renewed interest for these issues, state-lead land reform policies are no longer considered as a preferred option in international policy circles. Most of the land reforms implemented during the last 30 to 40 years were politically motivated and have not lived up to expectations. As Horowitz (1993) points out, instead of aiming to increase productivity and reduce poverty, land reforms often aimed at defusing social unrest and alleviating political pressures by peasant's organizations. Even where there was a genuine commitment at breaking-up the power of landed elites, agrarian reforms were generally designed by urban intellectuals with scarce knowledge of the realities of agricultural production and certain reservations about the potential efficiency of small-scale farmers (Barraclough, 1970). Finally, rather than improving the way land markets function and using such markets as a complement for governmental efforts to redistribute agricultural land, land reforms often provided substitutes for these markets, resulting in complex regulation systems that overstretched the available administrative capacity of the state (Lipton, 1974).

As Carter (1997) argues, the novel insights brought together by the new micro- and macro-economics of inequality have been twined instead with a sharp closure of development policy around a decidedly liberal orientation. Under this paradigm, property rights reforms - which assign legally secure and usually marketable land rights to individuals - and a constructive engagement with land markets, appear as main instruments for contemporary land policy.

Theoretical supporters of land titling and registration programs assert that well-established property rights and organized systems for public registration of property are an essential condition to improve the dynamics of land markets and move towards a more efficient distribution of resources. At the same time, poor households will then be able to use their secure assets as collateral for loans and will have a security-induced incentive to invest in their improvement, contributing in that way to increase the market value of their property and improve their competitiveness in the land market (Demsetz, 1967; De Alessi, 1980; Barzel, 1989; Libecap, 1989; Feder and Feeny, 1991).

From this standpoint, a more efficient use of land can be generated by two distinct sources. First, more efficient cropping choices are made possible because decision biases in favor of short-cycle crops that arise from tenure insecurity are removed with the introduction of land titles. Second, land is transferred from less to more dynamic farmers and consolidated into larger viable holdings, thereby eliminating the excessive fragmentation and subdivisions resulting

[1] For a complete review on the agrarian question debate and its application to Latin America, see de Janvry (1981), particularly Chapter 3.

from traditional land allocation and inheritance patterns. When property rights are not clearly ascertained nor effectively enforced, willing buyers who do not belong to the same community must incur into significant search, enforcement and litigation costs, as a result of which a gap is driven between the land's value according to the marginal product under the owner's use and the potential value of marginal product if used by the most productive alternative user. The price of land then does not reflect its true social value. Due to significant transaction costs arising from asymmetric information, land transfers are inhibited among unfamiliar farmers, thus causing the volume of land transactions to be less than optimal. By putting an end to ambiguity in property rights, land titling can drastically reduce transaction costs and encourage land acquisition by those able to make the best use of it (Platteau, 2000).

However, full-fledged private property rights do not only improve the allocation of land between different forms of uses and among different types of users, but also enhance investment incentives. Landowners whose rights are legally protected can be expected to be both more willing and more able to undertake investments. Their willingness to invest is enhanced via two channels. First, when farmers are better assured of reaping the future benefits of their present effort, thanks to secure rights of use, they have more incentives to invest in soil conservation measures, land improvements, and other operations that raise productivity in the long term.[2] The lack of tenure security can also be thought of as creating a risk of land loss that causes a decline of expected income from investments or, alternative, it may shorten the farmer's time horizon, thereby discouraging them from performing actions that increase benefits over time. A clear definition and registration of full-fledged private property rights is then supposed to provide land holders with the required level of tenure security and therefore increase their willingness to invest in the land (Demsetz, 1967; Feder *et al.*, 1988; Barzel, 1989; Libecap, 1989; Feder and Feeny, 1991; Besley, 1995; Binswanger *et al.*, 1995). Second, when superior transfer rights lower the costs of exchange in the land market, land becomes a more 'liquid' asset and hence any improvement made trough investments can be better realized. Investment incentives are then again enhanced (Besley, 1995; Platteau, 1996c).

Finally, farmers are not only willing but also more able to invest because the establishment of freehold titles increases the collateral value of land for credit lenders, principally by reducing their foreclosure cost in case of default, and allow farmers to receive better credit conditions to finance their investment projects (Feder *et al.*, 1988; Besley, 1995; Binswanger *et al.*, 1995). This is especially true regarding formal lending sources which often have imperfect information on the borrower.

However, so far the empirical evidence supporting these arguments is largely inconclusive. The presence of multiple market imperfections in these recently liberalized rural economies, and the subsistence of informal or customary property rights, seems to determine in practice whether or not these effects appear, their relative importance, and also its consequences in

[2] As Platteau (2000) points out, this fundamental idea can be traced back to the writings of John Stuart Mill (1848).

terms of efficiency as well as equity goals. Hence, in contexts where credit markets are missing or do not function properly, there may be little justification for this type of intervention when the lack of credit was thought to be the main limitation for investments (Platteau, 1996). If titling improves credit access but only for farmers with initially higher wealth levels (Zimmerman and Carter, 1999; Carter and Olinto, 2003), then the titling policy will raise concerns in terms of its distributional effect. If, on the other hand, the lack of tenure security is an important constraint for farmers to undertake investments, and titling helps to improve it, the policy may provide large benefits to the poor who are usually less able to acquire security by other informal means (Deininger and Chamorro, 2004). Moreover, if increasing levels of land formalization does not help to activate land sales or rental markets, not only the efficient allocation effect will be absent, but it can also set limits to the impact of titling on credit access and investment incentives.

1.2 Objectives and research questions

The general objective of this study is to improve our understanding of the relationship between property rights and investment incentives, with an emphasis on the potential impacts and limitations of Land Titling and Registration Programs. This means verifying whether land titling constitutes a necessary and/or sufficient condition to promote investments and increase land values. To accomplish this objective, we use information at two different levels. We assemble a country-level panel dataset for the macro perspective, and rely on household's surveys collected during the year 2004 as part of the evaluation of the PETT Program for the micro approach of this study. Our main research questions are the following:
1. What is the role of property rights in shaping the relationship between land distribution and economic growth?
2. How do legal documents affect farmer's tenure security and land-related investments? Is land titling required to enhance this effect?
3. What are the principal determinants and constraints that farmers face for accessing to formal sources of credit? Can land titling lift up some of these impediments and improve credit access for its beneficiaries?
4. Can land titling programs generate an externality effect on investments and land values by increasing the regional coverage of land rights formalization?

To address these four questions we will use three different levels of analysis. Our first research question is strictly explored at a macro level. A micro perspective is then used to analyze the relationship between property rights and tenure security, and finally, a meso approach is also included when investigating the links between land titling, credit access, and investments.

1.3 Analytical framework

Our four research questions will be addressed within the analytical framework shown in Figure 1.1, which provides an overall picture of the relationships explored in this study, and allows

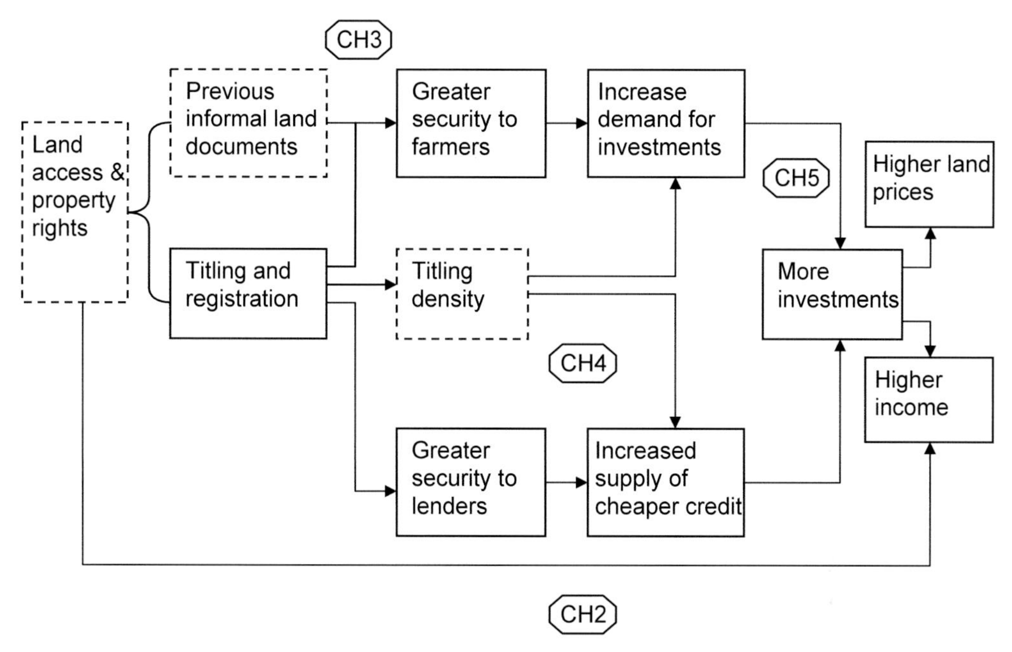

Figure 1.1. Analytical framework.

us to analyze each question in a specific chapter. The figure is based on the main relationships discussed in the seminal work of Feder *et al.* (1988), illustrated in solid-line boxes, and it also includes the new mechanisms and mediating variables that constitute the core of this study, depicted in dash-line boxes.

This analytical framework focuses attention on four specific aspects of land rights in relationship to development. The underlying logic of each of these features, which draws from the description presented in section 1.1, can be described as follows:

1.3.1 Land access, property rights, and economic growth

The relationship between land distribution, property rights, and economic growth has been rather unexplored at the macroeconomic level. Even though many studies address separately the implications of wealth inequality for economic growth, on the one side, and the importance of institutions like a well-defined property rights system on the other side, so far there is no study that jointly explores these links. Moreover, most studies exploring the relationship between inequality and growth rely on measures of income inequality rather than assets distribution as an explanatory variable. This is troublesome since the theoretical relationship between inequality and growth is better explained by assets distribution than by income.

Not only assets inequality can create a negative impact on growth, but also the degree of tenure security over those assets and the property rights system in a country will influence investments and growth. Studies that address the relationship between institutions and growth (Rodrik, 2000) argue that secure and stable property rights are a key incentive to invest.[3] A clear specification of land rights can also play a crucial role in correcting financial market imperfections, given their collateral function.[4] Moreover, securing land property rights would be key to reduce transaction costs in factor markets and thus improve economic efficiency and economic growth (Lipton, 1974; Byamugisha, 1999; Deininger *et al.*, 2003).

This study provides a theoretical discussion and empirical evidence to better understand the relationships between land distribution and economic growth when the role of secure property rights is accounted for. We assembled a new panel database that includes observations for more than 30 countries over the last three decades. The data include a time-varying variable for changes in the Land Gini index over this period that enables to overcome the limitations of previous studies that only included time-invariant measurement.

1.3.2 Land rights formalization, tenure security, and investment' incentives

The lack of tenure security over land is widely recognized as an important limitation for farmers to maximize the potential returns of this resource. Unclear definition of individual property rights can give rise to disputes over ownership, inheritance, or land boundaries. Ultimately, it increases the probability of losing the land in a dispute, and with it all effort and investments spent on it.

Even though land titling programs are fundamentally promoted to increase farmer's tenure security and investment incentives, the justification for this type of public intervention is increasingly questioned on the grounds of its limitation to replace or improve the effect of informal or customary rights already in place. Some authors argue that in many rural areas, customary rights provided by local authorities, or farmer's acquisition of informal land documents, might be sufficient to provide them with the required tenure security to induce investments (Migot Adholla, 1991; Platteau, 1992; Bruce and Migot-Adholla, 1994).

In areas where individual land ownership predominates, land titling programs concentrated on the formalization of previous land rights, with procedures based on the recognition of informal documents and reliance on community rules, could in fact contribute substantially to increasing farmer's tenure security and therefore boost land investments. Moreover, if the possibility of acquiring 'stronger' informal documents is related to wealth characteristics of farmers, titling might have a justification not only from an efficiency point of view but also from an equity perspective.

[3] The World Bank report by Deininger and Binswanger (1999) provides a comprehensive summary.

[4] Various studies have found that although land titles are a *necessary* condition they are not *sufficient* for getting access to (formal sources of) credit. See for example Boucher *et al.* (2004) and Van Tassel (2004).

1.3.3 Credit constraints in the agricultural sector

The provision and registration of land titles has been hypothesized to have a direct impact on farmer's access to credit because of its effect on increasing the collateral value of land for credit lenders. This effect would be especially true regarding formal credit sources which often have imperfect information on borrowers and thus insist on collateral before advancing a loan. However, a large amount of evidence suggest a weak or even null impact of titling programs on credit access, particularly in Latin America (Boucher *et al.*, 2004; Guirkinger and Boucher, 2006). In addition, in the few cases where a positive effect could indeed be established it was found to be mostly in favor of wealthier producers (Aldana and Fort, 2001; Carter and Olinto, 2003).

This study explores the characteristics of supply and demand for formal loans in the Peruvian agricultural sector, and analyzes the principal determinants and constraints that farmers face for accessing these sources of credit. Special attention is placed on the potential effect of the Peruvian Titling Program for lifting up some of these impediments and increasing credit access for its beneficiaries. Based on these results we identify complementary policies required for improving farmers, and particularly small-farmers, access to financial resources.

1.3.4 The externality effect of titling on investments and land values

Land titling programs are usually based on the assumption that full-fledged land rights provide incentives and opportunities for individual farmers to invest in improved resource-use strategies. Current approaches devote little attention to the importance of a scale of titling and to the potential role of externalities for the development of local factor markets. This study, therefore, explores the implications of a new possible impact of titling on land investments and land values derived from an 'externality effect' that emerges with an increase on the number of titled plots in the same district (titling density).

The potential effects of land titling on the willingness as well as on the ability of farmers to invest in their land are closely related to the functioning of other markets, most notably on the markets for land sales and credit (Feder and Feeny, 1991; Binswanger, Deininger *et al.*, 1995; Deininger and Binswanger, 1999). Even though titling is supposed to facilitate land transactions by reducing the cost of exchange on the land market, and to improve credit access by giving land a collateral value, several studies find that individual titling does not seem to be a sufficient condition for these markets to develop or work properly (Collier, 1983; Carter *et al.*, 1994; Lopez, 1996; Carter and Olinto, 2003; Boucher, Barham *et al.*, 2004).

We believe that one important condition for improving the functioning of these markets that has been overlooked in the literature, has to do with the need to count with sufficient density of formalized land rights in the area where parcels are located. By reducing overall transaction costs, titling density might help to improve the dynamics of land markets and affects investment's incentives via two different channels. On the one hand, if land markets expand as a result of an increasing general level of tenure security in the region, land becomes a more 'liquid' asset and hence any improvement made through investments can be easier realized when land is transacted. On the other hand, formal financial institutions could be more willing to locate themselves and provide loans in areas with a higher percentage of titled plots, since it will probably be easier for them to capitalize the land given as collateral in case of defaults on loans. If credit is required for making land investments, and one of the major limitations that farmers meet came from the supply side, we could expect an improvement in their ability to make investments when titling density in the area increases. The primary aim of this analysis is thus to explore the effect of titling on investments and land values while taking into account the general level of formalization of the land rights in the districts where the parcels are located.

1.4 Relevance of the study

The present study contributes to the existing literature in a number of ways, based on an analysis of the different mechanisms by which well-establish property rights can influence farmer's investment decisions.

First, this study adds to the literature on the relationship between inequality and economic growth in two different aspects. It replaces the commonly used measure of income inequality by a dynamic one of assets inequality (land Gini) which better reflects the arguments of most theoretical models. In addition, it incorporates new arguments that link inequality of assets and property rights institutions with economic growth, contributing to the discussion on the potential effects of redistributive policies as well as complementary interventions to guarantee its correct functioning.

Second, this study analyses the effect of titling on tenure security and investments while taking into account previous levels of tenure security provided by the earlier acquisition of different types of informal land documents. Because titling might have a differentiated effect on investments - depending on the initial level of tenure security - and since the acquisition of 'stronger' documents might be limited for poorer farmers, previous studies that do not account for these features are likely to overlook important consequences of this policy.

Finally, this study adds a new dimension to the analysis of land titling policies by introducing the notion of 'titling density' in our analytical framework. Current approaches devote little attention to the importance of scale in titling and to the potential role of externalities for the development of local factor markets. This study therefore explores the implications of a new possible impact of titling on land investments and land values derived from an 'externality effect' that emerges with an increase on the number of titled plots in the same district (titling density). The insights gained from such analysis may call for the introduction of a new regional perspective in the promotion of land titling programs and complementary policies to improve the livelihoods of the rural poor.

1.5 Outline of the thesis

Chapter 2 provides a theoretical discussion and empirical evidence to better understand the relationships between land distribution and economic growth when the role of secure property rights is accounted for. We assembled a new panel database that includes observations for more than 30 countries over the last three decades. The data include a time-varying variable for changes in the Land Gini index over this period that enables to overcome the limitations of previous studies that only included a time-invariant measurement. A system General Method of Moments (GMM) estimator is used to generate unbiased and consistent estimates for the parameters of interest.

The empirical analysis in the remaining chapters is based on information from household's surveys collected during the year 2004 as part of the evaluation of the Peruvian Land Titling and Registration Program (PETT). This evaluation was implemented by a research team at the Group of Analysis for Development-GRADE with our collaboration. The methodology implemented by PETT during the titling and registration process (T&R) is one of 'universal coverage' and free of any direct cost for the participants, thus in principle ruling out any potential self-selection bias. The survey was conducted amongst more than 2,000 farmers distributed over five different regional domains in the Coastal and Andean region of Peru. The sample frame used for this study was the National Cadastre Database, with information on more than 2 million parcels at the national level. As a parcel is added to this database when the T&R process starts, this means that in principle all parcels in our sample are potential beneficiaries of the program. In 2004, the year when the design was implemented, some of these parcels had already received a registered title while others did not yet received it. For the purpose of this study, we make use of the information on the year that a parcel was subject to T&R in order to divide them into 'treatment' and 'controls' groups. Treated parcels are the ones under T&R by the program during its first stage (1994-2000), while the control group is conformed by parcels Not-T&R at the time of the survey.

In Chapter 3 we use information on the type of informal document that the parcels had before the start of the program to explore their initial levels of tenure security and investments, as well as the characteristics of farmers that were able to acquire stronger documents. The effect of titling on investments is then analyzed for parcels with different initial levels of tenure security using a difference-in-difference estimation technique. We take advantage of having recall-data on all land-related investments made in each parcel, and the year that they were made, in order to apply this method.

Chapter 4 explores the characteristics of supply and demand for formal loans in the Peruvian agricultural sector, and analyzes the principal determinants and constraints that farmers face for accessing these sources of credit. Special attention is placed on the potential effect of the Peruvian Titling Program on lifting up some of these impediments and increasing credit access for its beneficiaries. We use survey questions specifically designed to identify rationing mechanisms for each individual, and a multinomial logit regression to determine the probability of being in each of them.

Chapter 5 investigates the effect of titling on investments and land values when taking into account the general level of land rights formalization in the districts where the parcels are located. We combine information from the survey with district level data from the National Agrarian Census of 1994, and test whether investment incentives are enhanced for parcels located in districts with higher levels of titling density. Thereafter we examine if individual titling and the level of titling density do only affect the perception of land prices via the investment effect, or do they also contribute in an independent way by reducing private enforcement costs and expanding market exchange opportunities.

Chapter 6 discusses the main findings of these studies and places them into a wider theoretical and policy perspective. Innovations, shortcomings, and topics for future research are also mentioned here.

2. Land inequality and economic growth: a dynamic panel data approach[5]

Abstract

The growing body of literature devoted to study the impact of inequality on economic growth have centred its attention on the income distribution effect, even though the theoretical relationships are more related to assets distribution. While some recent studies have tried to overcome this limitation by introducing asset indicators, they meet new constraints when dealing only with time-invariant measurements for this explanatory variable. This article provides a theoretical discussion and some novel empirical tests to better understand the relationships between assets distribution and economic growth. We assembled a new panel database that includes observations for more than 30 countries over the last three decades. The data include a time-varying variable for changes in the Land Gini index over this period that enables to overcome the limitations of previous studies. A system General Method of Moments (GMM) estimator is used to generate truly unbiased and consistent estimates for the parameters of interest. We explore some of the likely channels through which asset distribution and economic growth may be linked, paying particular attention to the role of secure property rights and the relations between land ownership and education. We find robust and significant negative signs for land inequality in the growth regressions, indicating that changes in asset distribution are an important factor for economic development.

2.1 Introduction

The relationship between inequality and economic growth has been subject to considerable debate in development circles. Following the seminal work of Kuznets (1955), a growing empirical literature addresses the linkages between income inequality and economic growth. While most studies consider inequality detrimental for growth, some recent findings point towards a possible positive relationship. Such differences are strongly determined by the type of indicators and the estimation procedures that are used.

Almost all studies that explore the relationship between inequality and growth rely on measures of income inequality rather than asset distribution as an explanatory variable. This is troublesome since the theoretical relationship between inequality and growth is better explained by assets distribution than by income. Inequality in assets is likely to reduce growth prospects due to insecure property rights or social polarization that reduce investment prospects. Asset redistribution in the form of land or education reforms can also play an important role in improving growth performance. For policy purposes, it makes a large difference whether

[5] This chapter is based on: R. Fort (2007), Land Inequality and Economic Growth: A Dynamic Panel Data Approach, accepted for publication in *Agricultural Economics*. An earlier version of the paper has been presented at the XXVI IAAE Conference in Gold Coast, Australia (12-18 August 2006).

inequality of income or inequality of assets is the underlying factor of registered differences in economic growth.

Most current studies rely on cross-sectional evidence rather than on panel data analysis and may thus provide biased results. The results obtained from this data can hardly be considered as adequate structural estimates, given the presence of country-specific attributes such as initial factor endowments or the country's particular history. Moreover, when panel data have been used to test the relationship between income inequality and growth, sometimes the traditional negative effect disappears, thus giving policy makers an argument to focus on growth-enhancing policies without worrying about distributional issues.

This chapter provides a theoretical discussion and some novel empirical tests to understand the relationships between assets distribution and economic growth. We explore the channels through which these processes are linked, paying particular attention to the role of human capital. In addition to traditional approaches that refer to credit market imperfections and investment constraints, we incorporate some new arguments that link inequality of assets with delayed growth through weak property rights institutions.

In order to avoid the common methodological problems stated before, we assembled a new panel database that includes observations for more than 30 countries over the last three decades. The data include a time-varying variable for the Land Gini index over this period that enables us to overcome the limitations of previous studies that included only a time-invariant measurement. A system GMM estimator as proposed by Arellano and Bover (1995) is used to generate truly unbiased and consistent estimates for the parameters of interest.

The chapter is structured as follows. First we present a review of the different theoretical models that explain the implications of asset inequality for economic growth. Hereafter, we outline the econometric estimation procedure. Next, we discuss the results of our estimations. We conclude with some implications for policy and further research.

2.2 Inequality and growth

The recent literature on the relationship between inequality and growth distinguishes two broad types of approaches that focus on particular channels through which these processes are linked. Following Dominicis *et al.* (2006) we refer to these approaches as the 'political economy' models and the 'socio-political instability' models. In political economy models, inequality affects taxation through the political process by which individuals can choose the tax rate directly or via electing governments that promote certain redistributive policies (Alesina and Rodrik, 1994; Persson and Tabellini, 1994). In very unequal societies we would expect then that more voters will prefer larger redistribution. If redistribution reduces the incentives to invest, and hence the growth rate, it is to expect then that more unequal societies will grow slower.[6]

More extended political economy models with capital market imperfections include credit constraints that prevent the poor from undertaking profitable investments. A more egalitarian wealth distribution can help to overcome asset thresholds and might result in higher aggregate investment in physical or human capital. As Stiglitz (1969) pointed out, when there are decreasing returns to capital and capital markets are imperfect, aggregate level of output may be affected by the wealth distribution. Aghion *et al.* (1999) used an endogenous growth model where redistributing wealth from the rich to the poor (whose marginal productivity of investment is relatively high) increases aggregate productivity and therefore growth. Under such conditions, asset redistribution creates investment opportunities in the absence of well-functioning capital markets, which in turn will enhance aggregate productivity and growth.

Socio-political instability approaches devote more attention to the role of social stability and property rights. Through its impact on economic efficiency, the distribution of assets can affect the cost of market exchange, the incentives to invest, the levels of violence, and the societies' ability to respond to exogenous shocks (Deininger and Olinto, 1999). Inequality can also create barriers that affect the cost of social interaction and economic exchange (Collier, 1998; Temple, 1998). Finally, inequality can be associated with violence and crime which will affect growth through the direct damage, the need to spend resources on preventive measures, and the impact on property rights and investment incentives (Knack and Keefer, 1995; Bourguignon, 1998).

In a recent study, Keefer and Knack (2002) argue that social polarization – measured by the inequality of land holdings – affects the likelihood of extreme policy deviations, making property rights less secure and thus negatively affecting growth. Once controlling for this indirect effect of inequality on growth, the direct link is likely to diminish. In a similar vein, it can be argued that not only inequality of assets can create a negative impact in growth, but also the degree of tenure (in)security over those assets and the property right system in a country will influence investments and growth. Studies that address the relationship between institutions and growth (Rodrik, 2000) argue that secure and stable property rights are a key incentive to invest.[7] A clear specification of land rights also plays a crucial role in correcting financial market imperfections, given their collateral function.[8] Moreover, securing land

[6] Although these models account for a negative correlation between inequality and growth, its mechanism does not seems to be supported by the data because some empirical studies found a positive rather than a negative effect of redistribution on growth (Easterly and Rebelo, 1993). When redistribution measures such as the tax rate or the level of social spending are regressed on measures of inequality, the coefficients are either insignificant or have a sign opposite to what theory predicts (Perotti 1996; Lindert 1996).

[7] The World Bank report by Deininger and Binswanger (1999) provides a comprehensive summary.

[8] Various studies have found that although land titles are a *necessary* condition they are not *sufficient* for getting access to (formal sources of) credit. See for example Boucher *et al.* (2004) and Van Tassel (2004).

property rights will be a key to reduce transaction cost in factor markets and thus improve economic efficiency and economic growth (Lipton, 1974; Byamugisha, 1999; Deininger *et al.*, 2003).

Most studies that analyze the systematic relationship between inequality and growth are based on rather simple measures of income inequality. Using cross-country data they find a negative relationship between income inequality and growth (Alesina and Rodrik, 1994; Persson and Tabellini, 1994). When a variable for initial land inequality is included, it is usually negatively associated to growth. An important extension to these approaches would be to examine the dynamic relationship between changes in asset distribution and economic growth.

Recently, new and larger data sets have become available that allow the incorporation of more sophisticated panel techniques. Studies by Forbes (2000) and Li and Zou (1998) using fixed-effects estimators to control for country-specific characteristics and dynamic GMM estimators to correct for endogeneity suggest that the negative relationship between inequality and growth weakens considerably and may actually be reversed.[9]

Various studies reviewed the consistency of these results using different specifications of income inequality (i.e. Gini coefficients, quintile shares, income ratios), different country samples and time periods.[10] Dominicis *et al.* (2006) used meta-analysis procedures to review existing evidence from 21 studies and conclude that inequality affects growth in a different way in higher and less developed countries.

A number of these recent contributions examine the possibility that - in line with the theoretical models discussed above - it is not so much inequality of income but unequal distribution of assets that may cause the reduction in countries' growth rates (Deininger and Squire, 1998; Birdsall and Londoño, 1998; Keefer and Knack, 2002). However, empirical evidence has been largely based on cross-sectional country level data rather than panel data analysis. Due to differences in the variables used (income vs. assets distribution) or the methods applied (cross-section vs. panel data), the empirical literature showed ambiguous predictions regarding the possible impact of inequality on growth.

The study conducted by Deininger and Olinto (1999) is an attempt to overcome these limitations by putting together a comprehensive panel data set with asset inequality between countries. They use a GMM estimator approach to examine the robustness of the inequality-growth relationship, including Gini coefficients to account for the initial land distribution (for the period 1960-70). This study not only identifies a significant negative impact of income

[9] Such positive effects of inequality on growth might be explained by the higher savings amongst rich households and the possibilities to overcome sunk costs in large investment projects.

[10] Galor and Moav (2004) provide evidence that the distribution-growth relationship depends on the development stage of a country, with more inequality at early stages of industrialization and more equality after people start to invest in education.

and asset inequality on growth rates, but also analyzes some of the channels through which this effect takes place.

We further explore what we consider two critical limitations of the former analysis. The first one is related to the use of a proper database, while the second one deals with potential gaps in the theoretical approach. While Deininger and Olinto (1999) recognize that inequality of assets is likely to be more stable inter-temporally than the distribution of income, they implicitly assume that asset distribution remains unchanged over a long time period. Moreover, it is less appropriate to use a time-invariant land Gini coefficient for the 1960-70 period as main variable, since many countries included in their sample made important land distribution reforms precisely right after that decade.[11] The collection of new information about land Gini distribution for several countries and different time periods allows us to analyze how changes in this variable - and not only their initial level – affect the relationship with growth.

The theoretical gap refers to the weak explanations that are usually offered regarding the potential direct effects of assets inequality on growth, controlling for the investment effect. It is argued that what could be behind this finding is either an 'incentive effect' or a 'social capital' effect, whereby inequality would increase the cost of social and economic interaction, including the ability to maintain the rule of the law in an unbiased way. Following this argument, and in line with the recently forwarded theoretical arguments regarding the linkages between property rights, inequality of assets, and growth, we need to test if - once controlling for the stability of property rights in each of the countries - the direct effect of assets inequality on growth is still maintained.

2.3 Econometric estimation and data specification

We start from the usual equation in the empirical analyses of the determinants of growth:

$$(y_{it} - y_{it-1}) = \alpha y_{it-1} + \beta' X_{it-1} + \delta' Z_i + \varepsilon_{it} \qquad (1)$$

where y_{it} denotes the logarithm of per-capita GDP of country i in period t, X_{it-1} is a vector of country-specific time-varying variables affecting growth, and Z_i is a vector of country specific time-invariant variables, and ε_{it} is an error term that captures the effect of time-invariant and time-varying unobserved country characteristics. The disturbance term ε_{it} can be divided in a country-specific time-invariant effect u_i and the time-variant disturbance e_{it}.

We assume that $Cov(e_{it}, u_i)=0$ and $Cov(e_{it}, e_{is})=0$, for any $t \neq s$. Eq. (1) becomes then:

$$(y_{it} - y_{it-1}) = \alpha y_{it-1} + \beta' X_{it-1} + \delta' Z_i + u_i + e_{it} \qquad (2)$$

[11] This is the case for example in many Latin American countries that present the most unequal distribution of land in the sample used for the Deininger & Olinto study.

The OLS estimation of the parameters in equation (2) is likely to be biased and inconsistent for two reasons: first, by construction y_{it-1} is correlated with the country-specific effect u_i, and second, it is likely that some of the variables in vectors X_{it-1} and Z_i are also correlated with that error component. Second, asset inequality can be correlated to factor endowments, and conditioned by the country-specific history which are unobservable characteristics measured by u_i.

The usual solution to this lack of orthogonality with panel data is to estimate the specified parameters by applying OLS to the 'within groups' transformation, or 'first differencing' left and right-hand-side variables in (2). In this particular case, however, estimation of equation (2) by 'fixed effects' would create some other problems. First, given the dynamic nature of the model, the first difference of y_{it-1}, defined as $\Delta y_{it-1} = y_{it-1} - y_{it-2}$ is by construction correlated to the first difference of the error component e_{it}, given by $\Delta e_{it} = e_{it} - e_{it-1}$. Second, even though X_{it-1} is uncorrelated by assumption to the error component e_{it}, X_{it} is likely to be contemporaneously correlated to e_{it}, which implies that ΔX_{it-1} will be correlated to Δe_{it}. Therefore, the OLS estimator of α and β obtained by regressing Δy_{it} on Δy_{it-1} and ΔX_{it-1} will be biased and inconsistent. Since the first difference of Z_i is zero, we would not be able to identify their parameters using fixed effect estimation methods.

The lack of identification for the time-invariant variables when the within transformation is adopted can be solved by employing the Instrumental Variables (IV) estimator proposed by Hausman and Taylor (1981). For the lack of orthogonality between Δy_{it-1}, ΔX_{it-1} and Δe_{it}, inherent to dynamic panel data models, Arellano and Bond (1991) formulate a consistent and unbiased GMM estimator which uses twice lagged y_{it} and X_{it} as instruments. An extension of this model proposed by Arellano and Bover (1995) provides a unifying GMM framework that can be generalized for the estimation of Hausman and Taylor type models, as well as dynamic panel data models. In addition of using instruments in levels for the equations in first differences, we also use instruments in first differences for the equation in levels, which allow us to estimate a 'system GMM' instead of only a 'differences GMM' model as the one proposed by Arellano and Bond (1991). With this addition we can estimate the parameters in the levels equation that explain a substantial part of the total variation in the data.

We composed a new data set on the distribution of operational holdings of agricultural land from the decennial FAO World Census of Agriculture and other sources for 35 countries. For each country, this information has been recovered for three different periods over time: 1960-1970, 1971-1980, and 1981-1990, giving us a total of 105 observations in the panel. We complement the data with measures of real GDP per capita and the share of investment in GDP from the Penn Word Tables data set, data on human capital stock taken from Barro and Lee (2000), and finally a time-invariant variable containing a measurement of the 'rule of law' for the 1980 decade taken from the ICRG (International Country Risk Guide) law and order rating.[12]

Time-varying information on the distribution of land holdings is scarce and particularly difficult to find for less-developed countries. It can be argued that the inclusion of several developed countries in the sample could be disturbing the analysis, because agricultural land is a less important asset for them than it is for the poorest ones. However, Carter (2000) shows a persistent relationship between land ownership inequality and income inequality over time, and provides some theoretical and empirical explanations for this link. Moreover, the inclusion of these countries, which in many cases experienced a strong agricultural transformation process to achieve their development goals, will work as a benchmark for the estimation of the level equations in our system.

The panel data (see Appendix 2.1) show that land Gini coefficients do not only vary across countries but also show considerable change over the studied time period. For example, the average Gini coefficient for the initial period is 0.60 and for the final period 0.62, both with standard deviations of 0.16. The average difference between the Gini of the initial period for all countries and that of the final period is 0.015 (standard deviation 0.06).

2.4 Results

The growth regression results are presented in Tables 2.1 and 2.2 for different specifications of the equation. As we can observe in the first column of Table 2.1, the coefficient for the land Gini distribution is negative and significant, thus confirming the hypothesis that both the level and the change towards a more equal distribution of land have an important positive effect on the GDP growth of a country.[13]

A very interesting result derived from column 3 is that once we include the interaction term between human capital and land distribution, the configuration of the inequality effect on growth changes dramatically. The coefficient for land distribution is now positive but insignificant, the education effect becomes significant, and the interaction term is negative and significant. This result gives support to the hypothesis that even though human capital investments are very important for enhancing growth, countries with highly unequal levels

[12] Since lagged variables are used as instruments for the estimation, the equation in levels for the earliest period (1970) as well as the last difference equation including periods (1980-1970) cannot be estimated. The 'System GMM' estimation of eq. (2) thus includes two equations for the regression in levels and other two for the regression in first-differences.

[13] This simple specification was also estimated without including the high income level countries in our sample, what maintained the sign and significance of the land gini coefficient. We also estimate this and all the other regressions with regional dummies, dummies classifying countries by income level (IMF rank), and time-varying measure of the percentage of land under cultivation (FAO-Stat). None of these inclusions led to a change in our main results.

Table 2.1. Growth regression with Land Gini and education.

	(1)	(2)	(3)
Initial GDP (log)	-0.014**	-0.011	-0.006*
	(0.005)	(0.010)	(0.003)
Land Gini	-0.121**	-0.104**	0.103
	(0.027)	(0.027)	(0.064)
Human capital (log)		0.012	0.066**
		(0.014)	(0.025)
Human cap.* Land Gini			-0.111**
			(0.044)
Intercept	0.222**	0.165**	0.017
	(0.054)	(0.079)	(0.043)
countries	33	31	31

**P<0.05; *P<0.1. Std errors between parentheses.

Table 2.2. Growth regression with land gini, education, investments, and rule of law.

	(1)	(2)	(3)	(4)
Initial GDP (log)	-0.006	-0.006	0.007	0.0086
	(0.004)	(0.006)	(0.012)	(0.0152)
Land Gini	-0.070**	-0.077**	-0.071**	-0.0569*
	(0.026)	(0.021)	(0.024)	(0.0308)
Initial investment (log)	0.023**	0.029**	0.009	0.0190*
	(0.010)	(0.008)	(0.007)	(0.0107)
Rule of Law index			0.055**	0.0577**
			(0.019)	(0.0236)
Rlaw * Initial GDP			-0.006**	-0.0064**
			(0.002)	(0.0029)
Human capital (log)		-0.007		0.0034
		(0.010)		(0.0113)
Intercept	0.044	0.040	-0.031	-0.0807
	(0.056)	(0.055)	(0.092)	(0.1304)
countries	33	31	29	29

**P<0.05; *P<0.1. Std errors between parentheses.

of asset distribution tend to face a reduced effectiveness of their educational policies. Putted in another way, we could argue that efforts for land redistribution should be implemented together with improvements in education in order to have a decisive impact on economic growth. This argument has been frequently forwarded in explaining the differences between land redistribution policies in Asian countries compared to the ones in Latin America (Birdsall *et al.*, 1995; Birdsall and Londono, 1998).

Column 1 in Table 2.2 examines whether there is an independent effect of the asset distribution once we include the investment variable in the model. As we can see, the investment coefficient is positive and significant as expected, but the land distribution coefficient remains significant and negatively related to the growth rate. However, its magnitude has been reduced with more than 40 percent compared to the model where it was the only regressor included. The independent effect of land distribution remains negative and significant even when human capital is added to the model (column 2).

Finally, we tested whether this apparent direct effect of inequality on growth is maintained once a measurement for the country's political stability is included. Some authors argue that the main effect of inequality on growth is through its indirect effect on the security or stability of property rights. This implies that once controlling for the Rule of Law, the coefficient of land distribution should decrease or even turn insignificant.[14] As the index for Rule of Law that we are using is highly correlated with the level of GDP (correlation of 0.7) we include it alone and also as an interaction with the initial level of GDP for each decade.[15] Column 3 shows that the index for Rule of Law has indeed a direct and positive impact on growth. While its addition to the equation turned the coefficient for the investment share insignificant, the direct negative impact of land distribution remains unaffected. The last column in Table 2.2 incorporates the human capital variable. Once we add this control the main effect is an important reduction in the coefficients for land distribution (from 0.07 in column 3 to 0.056) and for the investment share (from 0.029 in column 2 to 0.019), which is now again significant.[16]

2.5 Discussion

Using for the first time a panel data set with changes in land distribution over time and between countries we have been able to provide confirmation for the hypothesis that asset distribution is a major determinant of economic growth. Apart from a direct effect we also show that land inequality creates a barrier to the effectiveness of educational policies, confirming the

[14] Using a cross-section database Keefer and Knack (2002) find that, when an index for property rights is added to the growth equation, the land inequality coefficient is reduced in one third, even though still negative and significant.

[15] We also try some indicators from the IRIS International Country Risk data, like government repudiation of contracts, expropriations, and rule of law index, getting similar results as the ones presented here.

[16] The regression including all explanatory variables could not be run because of excessive number of instruments relative to the number of observations in the dataset.

initial findings of Deininger and Olinto (1999). Moreover, the incorporation of the physical investments variable in the model corroborate the existence of a growth reducing impact of land inequality that goes beyond the conventional channel of credit market imperfections and reduced investments.

Even though the security of property rights appears as an important factor to explain economic growth, its effect does not modify the relationship found between land inequality and growth, as Keefer and Knack (2002) argued. The omission of the investment variable in their model is the likely reason for this discrepancy. Future research needs to incorporate the potential relationship between property rights and investments in order to clarify their individual influence on the connection between land inequality and economic growth.

These results have two important implications for policy strategies. First, it becomes clear that policies aiming at a more equal distribution of assets will be more effective if combined with complementary measures towards educational reforms and the improvement of institutional arrangements to secure property rights. The lack of such a combined implementation of structural reforms can be one of the reasons why land reforms in several countries failed in the past to achieve the expected economic growth. Second, for developing countries that pursue market liberalization and privatization programs, it becomes of fundamental importance to remain alert that the effects of these reforms are not leading to the concentration of assets in few hands. Such unintended consequences are likely to deteriorate the country's economic performance in the long run.

In order to explore in more detail the conclusions derived from this study, some issues require further examination. It would be desirable to expand the sample of countries with accurate information about (changes in) land distribution, particularly to include more underdeveloped countries, so that more instruments and controls can be used in the analysis. Another option would be to obtain a broader measure of assets distribution (i.e. including housing and urban land ownership). It would also be important to find measures that are more directly reflecting ownership security. Many other factors - such as social interaction problems, political instability or ethnic heterogeneity - can be also playing a role but where not (yet) included in our analysis.

Appendix 2.1. Data base

Country	Land Gini			GDP per capita			Investment			Human capital			Rule
	1970	1980	1990	1970	1980	1990	1970	1980	1990	1970	1980	1990	Law
East Asia & Pacific													
FJI	0.65	0.84	0.77	2,592	3,609	3,985	0.19	0.24	0.12	4.9	5.8	6.4	-
IDN	0.55	0.55	0.46	715	1,281	1,974	0.11	0.18	0.28	2.5	3.1	2.9	1.9
JPN	0.47	0.52	0.59	7,307	10,072	14,331	0.40	0.34	0.39	5.1	5.4	5.5	5.0
KOR	0.37	0.35	0.34	1,680	3,093	6,673	0.22	0.28	0.37	3.5	4.8	5.5	2.3
PHL	0.51	0.51	0.55	1,403	1,879	1,763	0.13	0.19	0.18	3.7	4.7	5.0	1.0
THA	0.43	0.44	0.47	1,526	2,178	3,580	0.18	0.17	0.27	3.7	3.7	4.6	3.2
East Europe & C. Asia													
TUR	0.59	0.57	0.61	2,202	2,874	3,741	0.21	0.23	0.21	2.1	2.6	3.0	2.8
Latin America													
BRA	0.84	0.85	0.85	2,434	4,303	4,042	0.20	0.22	0.15	2.5	2.3	3.1	3.9
PAN	0.80	0.84	0.87	2,584	3,392	2,888	0.26	0.22	0.16	3.5	4.5	5.7	2.0
PER	0.92	0.91	0.86	2,736	2,875	2,188	0.13	0.23	0.16	3.4	4.2	4.1	1.0
PRI	0.79	0.77	0.77	5,780	6,924	8,972	0.32	0.16	-	-	-	-	-
PRY	0.86	0.93	0.93	1,394	2,534	2,128	0.09	0.21	0.18	3.4	4.0	4.4	2.0
Mid-East & N. Africa													
ISR	0.80	0.77	0.85	6,004	7,895	9,298	0.30	0.21	0.21	6.1	6.7	6.6	2.4
PRT	0.81	0.81	0.78	3,306	4,982	7,478	0.28	0.27	0.16	2.1	2.5	3.0	5.0
North America													
USA	0.72	0.72	0.74	12,963	15,295	18,054	0.20	0.20	0.20	5.8	5.9	5.8	5.8
South Asia													
IND	0.62	0.61	0.58	802	882	1,264	0.13	0.14	0.16	2.0	2.4	3.0	2.4
NPL	0.56	0.60	0.45	670	892	1,035	0.04	0.08	0.08	0.1	0.6	0.9	-
PAK	0.51	0.52	0.57	1,029	1,110	1,394	0.10	0.09	0.10	1.0	1.1	2.1	2.0
Western Europe													
AUT	0.70	0.69	0.65	7,510	10,509	12,695	0.28	0.28	0.26	3.6	3.7	3.6	6.0
BEL	0.60	0.58	0.56	8,331	11,109	13,232	0.27	0.24	0.25	6.8	5.9	6.0	6.0
CHE	0.51	0.52	0.50	12,942	14,301	16,505	0.31	0.30	0.35	5.1	5.5	5.4	5.8
CYP	0.62	0.61	0.63	3,753	5,295	8,368	0.32	0.30	0.23	4.1	4.4	5.4	2.5
DEU	0.51	0.52	0.68	9,425	11,920	14,341	0.32	0.27	0.26	3.6	3.7	3.7	-
DNK	0.43	0.46	0.44	9,670	11,342	13,909	0.30	0.22	0.21	5.5	5.5	5.5	5.9

Country	Land Gini			GDP per capita			Investment			Human capital			Rule
	1970	1980	1990	1970	1980	1990	1970	1980	1990	1970	1980	1990	Law
ESP	0.84	0.85	0.86	5,861	7,390	9,583	0.28	0.24	0.29	4.1	4.0	4.2	3.8
FIN	0.25	0.23	0.26	8,108	10,851	14,059	0.40	0.35	0.33	4.9	5.1	5.5	5.8
FRA	0.53	0.53	0.53	9,200	11,756	13,904	0.31	0.27	0.27	4.2	4.1	4.2	5.1
GBR	0.69	0.68	0.67	8,537	10,167	13,217	0.20	0.16	0.19	5.8	5.9	6.0	4.6
IRL	0.49	0.49	0.48	5,015	6,823	9,274	0.27	0.27	0.23	5.0	5.2	5.4	3.9
ITA	0.75	0.76	0.78	7,568	10,323	12,488	0.31	0.27	0.25	4.1	3.7	3.8	4.9
LUX	0.45	0.47	0.48	9,782	11,893	16,280	0.32	0.26	0.32	-	-	-	6.0
NLD	0.48	0.50	0.55	9,199	11,284	13,029	0.30	0.23	0.22	5.3	5.3	5.4	6.0
NOR	0.46	0.48	0.46	8,034	12,141	14,902	0.34	0.30	0.21	5.2	5.3	6.6	6.0

Land Gini: derived from FAO World Census of Agriculture (operational holdings).
GDP per capita: obtained from the Penn-Word Table 6.1
Investment: obtained from the Penn-Word Table 6.1.
Human Capital: taken from Barro and Lee (2000).
Rule Law: taken from the International Country Risk Guide (ICRG) database. Higher scores indicate 'sound political institutions, a strong court system, and provisions for an orderly succession of power'. Lower scores indicate 'a tradition of depending on physical force or illegal means to settle claims'. Index between 0-6.

3. The homogenization effect of land titling on investment incentives: evidence from Peru[17]

Abstract

Land titling programs have been widely promoted as a necessary condition to enhance farmer's incentives to invest in their land. The justification for public intervention of this type is increasingly questioned on the grounds of its limitation to replace or improve the effect of informal or customary rights already in place. When the aim of the program is concentrated on the formalization of previous land rights and its procedure is based on the recognition of informal documents and reliance on community rules, it could in fact contribute to increased farmer's tenure security and therefore boost land investments. We explore this relationship for a sample of Peruvian farmers that are part of a state-led land titling program which shares the aforementioned characteristics. Using retrospective information regarding the type of informal documents that parcels had before the start of the program we are able to categorize them into two different levels of initial tenure security. The effect of titling on investments is then analyzed for these two groups of parcels using a difference-in-difference estimation technique. Our results show that there is a positive effect of titling on the probability of making investments as well as on the value of investments for both groups of parcels, but also prove that its impact is higher for parcels with previously low levels of tenure security. Moreover, this effect can be almost entirely attributed to changes in farmer's willingness to invest and not to better access of credit. In conclusion, given that informal land rights constitute at best imperfect substitutes to full-fledged property rights, and that there seem to be many wealth-related limitations to acquire them, public provision of land titles appears to be a good option for enhancing farmer's willingness to invest in their land.

3.1 Introduction

Land Titling programs have been widely promoted as a necessary condition to enhance farmer's incentives to investments in their land, because of their potential effect on farmer's willingness and ability to make such investment efforts. Two different types of arguments are commonly used in this debate. On the one hand, the lack of tenure security can be thought of as creating a risk of land loss that causes the decline of expected income from investments or, alternative, it may shorten the farmer's time horizon, thereby discouraging them from performing actions that increase benefits over time. A clear definition and registration of full-fledged private property rights is then supposed to provide land holders with the required level of tenure security and therefore increase their willingness to invest on the land (Demsetz, 1967; Feder *et al.*, 1988; Barzel, 1989; Libecap, 1989; Feder and Feeny, 1991; Besley, 1995; Binswanger *et al.*, 1995). At the same time, the establishment of freehold titles increases the collateral value

[17] An earlier version of this chapter is under review in the NJAS-Wageningen Journal of Life Science.

of land for credit lenders by reducing their foreclosure cost in case of default, allowing farmers to receive better credit conditions to finance their investment projects (Carter *et al.*, 1994; Besley, 1995; Binswanger, Deininger *et al.*, 1995; Carter and Olinto, 2003).

Although there is little disagreement about the role of these factors at conceptual level, their relative importance in explaining the investment effect - and its consequences in terms of the distributional implications of land titling - have been subject to much debate in the literature. In settings where credit markets are missing or do not function well, there may be little justification for this type of intervention when the lack of credit was thought to be the main limitation for investments (Platteau, 1996). If titling improves credit access only for farmers that were already better-off (Zimmerman and Carter, 1999; Carter and Olinto, 2003), then the titling policy will raise concerns in terms of its distributional effect. If, on the other hand, the lack of tenure security is the principal constraint for farmers to undertake investments, and titling helps to improve it, the policy may provide large benefits to the poor who are usually less able to acquire security by other informal means (Deininger and Chamorro, 2004).

The need for a public intervention in the provision of titles with the intention to increase tenure security has, however, received many criticisms in the literature. A large part comes from studies on different African countries where titling policies proved to be ineffective for enhancing investments. The principal argument of these studies is that in customary land areas, basic land rights (i.e. freely choose which crop to grow, freely dispose of harvest output, prevent others from exploiting the same parcel) provided by local authorities or custom seem to be sufficient to induce land holders to invest, and that adding transfer rights (assumed to be brought by titling) does not appear to significantly improve investment incentives. Apparently, the local informal order embedded in rural communities of these areas guarantees basic land rights to all villagers which are sufficient to induce investments. In this situation, there will be no need for the state to intervene trough centralized procedures aimed at formalizing land rights (Atwood, 1990; Migot Adholla and *et al.*, 1991; Platteau, 1992; Bruce and Migot-Adholla, 1994; Platteau, 1996).

Even though property rights regimes[18] may differ between African and Latin American rural societies, some researchers have started to transmit these concerns about the relationship between customary or informal rights and the introduction of full-fledged private property rights to the debate on the latter region. According to Zoomers and van der Haar (2000), this interplay constitutes one of the most important issues that require further investigation to better understand the current land tenure situation in Latin America., Most studies that attempt to measure the effects of Land Titling policies in Latin America disregard informal land rights that are currently in place. One of the reasons for not considering them could be related to the fact that titling policies in the region have been mostly oriented to the 'formalization' of individual rights over pieces of land that were already privately hold, and

[18] Following Bromley (1998), property rights regimes comprise the nature of ownership, the rights and duties of the owners, the rules of use, and the locus of controls.

not so much to the process of 'privatization' of land held before under other types of property regimes.[19] Even in the case where the principal scope of the policy is the formalization of individual rights, it seems far too simplistic to assume that there were either no informal rights governing the rules of use and exchange of land before the tilting policy took place, or that these rights were 'homogeneous' amongst all plots and farmers such that the levels of tenure security before titling were all the same.

Relaxing these assumptions compels us to explore the different ways under which farmers build and enforce their private rights over the land, and to observe if these different arrangements result in heterogeneous levels of tenure security and investments between parcels. In particular, we want to know if there is a correlation between selected indicators of household's wealth and market integration and the probability of holding a document that provides higher levels of tenure security, and also if parcels with these types of documents presented higher investment's level prior to the start of the program. If this is the case, it could be expected that the effect of the titling program on tenure security and investments, if any, will depend on that initial level. A higher effect on investments for parcels with previous lower levels of tenure security would indicate a justification for the program, not only from an efficiency point of view but also from an equity perspective.

This chapter explores these hypotheses by using information from the Peruvian Land Titling and Registration Program. The particular history of land distribution in the country, as well as the characteristics of the program's implementation process and the performance of the sample of farmers selected for this study, makes this case particularly interesting and appropriate for addressing our research questions. The next section contains an overview of the changes in land policies in Peru during the last decades, and presents key aspect of the Land Titling program under analysis. Section 3 describes the database used for this study and presents a classification of parcels by tenure status before the start of the program. Section 4 formalizes the model and derives testable hypothesis. Section 5 deals with the econometric model to be estimated and confronts some of its potential problems. The estimation results are presented in Section 6 followed by some concluding remarks.

3.2 Land tenure reform in Peru

During the last three decades the legal framework regarding land issues in Peru has radically changed from a strongly regulated process towards a more market-based perspective. The Agrarian Reform Law in 1969 that sets the base line for a large transformation in the agrarian structure was followed by many restrictive laws about the use of land. A cooperative land ownership scheme was installed after the expropriation of the large haciendas from their prior owners and the new legislation included a prohibition to sell land received during this

[19] By 'privatization' we mean the change form a communal to a private ownership of a piece of land, and the consequent assignment of rights at the individual level. This case could be thought of as creating more conflicts between previous customary rights and the new individual rights brought by titling.

process. By the end of the seventies, most of these cooperatives went bankrupt and many farmers and their organizations initiate movements to push the government for a change in the law. Legislative Decree N.85 of 1981 established the possibility of dissolution for agricultural cooperatives, trying to promote a change in their management, but turned out to be the beginning of the fragmentation of many of them, transferring small pieces of land to their members.[20] Most of the times, these transfers from cooperatives to individual members did not encompass a property title issued by the state but only an informal document provided by the ex-cooperative or in some cases no document at all.

During the 1990s, Peru turned towards a more liberal regime in terms of land ownership and use of land. In 1991, Fujimori's government passed the Laws 653 and 667 which promoted cadastre and titling policies for rural areas as well as lifted some of the previous restrictions on land sales, rentals and mortgages. As Zegarra (1999) points out, only 10 percent of the total estimated number of parcels on 1990 was registered in a Registration Office. From this moment onwards, the definition of private property and the demands for well-defined property rights over the land acquired greater importance.

The '*Programa Especial de Titulación de Tierras*' (PETT Program) was created in 1992 in order to promote land titling and improve the situation of many farmers with different types of informal documents that supported their land ownership status. The program has a nation-wide perspective with the objective of constructing a rural cadastre system with validity all over the country. By the end of 2005, the PETT Program managed to title and register more than 1.5 million parcels, changing the percentage of formally-owned plots to more than 50 percent. In the last seven years the program budget accounts to more than 100 million dollars, what makes it also one of the largest formalization programs for rural areas in the developing world.[21]

Figure 3.1 shows the different steps followed by the program in order to award a registered title.[22] The methodology implemented by PETT during the titling and registration process (T&R) is one of 'universal coverage', which in principle rules out any potential self-selection bias of program participants. The program works in a strongly decentralized way, with several regional offices sending their personnel to the field simultaneously. The first step in the process is to create a cadastre-database of all the parcels over a certain region (normally a valley). They rely on aero-photography to contrast the pictures with the parcel information collected later on in the field together with the possessor and bordering neighbors (Bordering). Information about the possessor, field characteristics, and proof of informal rights over the land, are also collected at this stage (Census). Based on this information the PETT regional office produces

[20] Depending on the region, the type of production, and the status of the member, they got between 2 and 5 hectares of land.

[21] Sources: The Peruvian Ministry of Finance, The Inter-American Development Bank, and The World Bank.

[22] This information was collected during personal interviews with PETT officials and is also available on the program web page (www.pett.gob.pe).

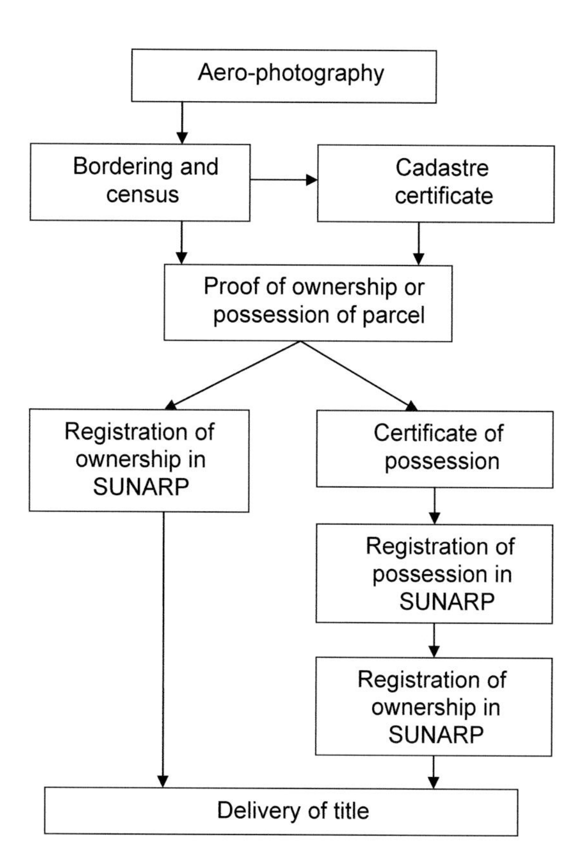

Figure 3.1. The titling and registration process under the PETT program

then a Cadastre Certificate, which will be required for all the titling process (Cadastre Certificate).

The next step in the process consists in registering the ownership rights in the public registration system (SUNARP).[23] For that matter PETT officials use the information recovered in the Census to distinguish two types of tenure regimes over the parcels (Proof of Ownership and Possession). If the documents presented by the farmer to the PETT officials provide enough evidence of ownership then the complete file for that parcel is ready to be send to the SUNARP. These documents include previous titles issued by the Agricultural Ministry, private transfer contracts certified by a notary, or judicial resolutions. If the farmer could not present any document to prove his/her rights over the parcel or had only other documents not considered as proof of ownership, the procedure consists in issuing first a Certificate of Possession, and transform it later on into an Ownership Certificate to be registered in the SUNARP.

[23] SUNARP is the 'Super Intendencia Nacional de Registrios Publicos'. All this procedure is regulated under Legislative Decree (LD) 667 and its posterior modifications in LD 889 and Law Decrees 26838 and 27161.

In order to get a Certificate of Possession, the farmer must prove direct, continuous, peaceful, and public possession of the parcel for a minimum of one year for state land or five years for private land. To do so, it is enough to present a written declaration of all adjacent neighbors or a declaration of the producers association in the region. Adding any other document that shows possession is recommended but not mandatory. This Certificate, together with the Cadastre Certificate, is then send to SUNARP to register the possession right of the farmer. If accepted, SUNARP will then notify the possessor and all community neighbors about the registration and give them 30 days to issue any complain. Passed that time, if no complain have been made, SUNARP will proceed to the registration of the ownership rights. The final step in the process is the delivery of the registered titles to the owners, which is commonly done in a massive way with a public ceremony.[24]

3.3 The classification of sampled parcels by tenure status

The database used for this study was collected during the last months of 2004 as part of the socio-economic evaluation process of the PETT program. The survey was conducted amongst more than 2,000 farmers distributed over five different regional domains on the Coastal and Andean region of Peru.[25] The sample frame used for this study was the National Cadastre Database, with information on more than 2 million parcels at the national level. As mentioned in the previous section, a parcel is added to this database when it just started the T&R process, which means that in principle all of them are potential beneficiaries of the program. At 2004, the year when the design was implemented, some of these parcels had already received a registered title, while others had not received it yet. This difference was the main feature for the initial selection of parcels as 'treatments' or 'controls'.

The survey also recovers recall-information on some variables that are of particular interest for this study. For example, data were gathered for different types of land-attached investments made in each parcel of the household, recording also the year that they were made. It terms of the tenure status of each parcel, we collected information on the type of document that they currently hold as a proof of ownership or possession, and the year in which they received it, as well as on the type of document that they had previous to that one. This information helps us to reconstruct the changes in tenure status for each parcel of the household throughout the years, and in particular to identify the previous status of the parcels before being T&R by the PETT program.

[24] Even though there could be a difference in the time it takes to register a parcel with a document that shows ownership as compared with one with a document that only shows possession or no document at all, delivering of titles is most of the times done in a massive ceremony where all titles from the same area are delivered together. We assume that a parcel is T&R from the moment the farmer obtains the titling document and not from the time of registration in SUNARP.

[25] The coastal region was divided between North-Coast and Center-South-Coast domains, and the Andean region between North-Andean, Center-Andean, and South-Andean domains.

We have seen that the PETT program makes in fact a distinction between the different types of informal documents during the T&R process. Some documents contain enough initial evidence of ownership, while others first need to be 'validated' by the rest of the community' members and by the Registration Office. This issue is an implicit recognition of previous land rights and might also be related to different levels of tenure security before titling. Even if these informal documents do not provide farmers with a complete 'bundle of rights' over their land, in some cases they can be sufficient to give farmers the tenure security needed to reduce their perception about any risk of losing the land in a dispute. If disputes over the land are mostly of local nature and can normally be solved by local authorities, it is likely that some of these documents provide enough enforcement power to make farmers feel secure. Apart from this when land transactions occur mostly between community members, some of these documents could provide enough security to the buyer in case of sales, or to the owner of the land in case of renting-out land for a period of time.

With this classification in mind, we make a further distinction in our treatment and control groups. Within the controls (without T&R) we divided parcels into 'low tenure security' (LTS) and 'medium tenure security' (MTS) according to the type of document that they currently hold. Within the treatment group (with T&R) we make use of the retrospective information on the type of document they had before getting the PETT title to subdivide them in a similar way. Table 3.1 shows the distribution of parcels by these groups and according to regional domains.

Parcels categorized under 'low tenure security' (LTS) are the ones without any document to prove possession or ownership, or the ones with a possession certificate of any Agricultural Ministry' Local Agency or a Peasants Community, or a certificate of having register a piece of public/abandoned land under your name. Parcels categorized under 'medium tenure security' (MTS) are the ones that count with old titles issued by the Agricultural Ministry, a buy-sell contract, or some type of public deed certified by a local judge or notary. The idea behind

Table 3.1. Distribution of parcels by domain and groups.

	MTS		**LTS**		**Total**
	Treatment	**Control**	**Treatment**	**Control**	
North Coast	69	71	148	93	381
Center-South Coast	60	35	39	17	151
North Andean	116	191	25	130	462
Center Andean	220	227	148	128	723
South Andean	116	227	24	146	513
Total	581	751	384	514	2,230

this categorization of parcels according to the type of documents is that land property rights in this setting can be better understood as a 'continuum of rights' instead of just a discrete indicator.[26] In this sense, low tenure security documents provide inferior rights as they serve at most to prove possession of a parcel, but they cannot be used as a proof of ownership. Medium tenure security documents can be legally used to proof ownership of a parcel but they lack of 'universal' recognition and approval as they are not registered in the public system.[27] In this continuum, registered titles provided by the PETT program are supposed to give farmers the highest tenure security over their parcels.

One concern related to the validity of this classification has to do with the possible relationship between the length of possession of the parcel and the type of informal document that the farmer holds. If the decision to acquire a MTS document comes mostly after certain years of working on the plot, and perhaps only after having made some investments on it, then length of possession will probably be the most relevant variable to differentiate parcels into tenure security levels previous to tilting. To show that this does not seem to be the case in our study, we first look at the distribution of the years of possession for parcels in the LTS and MTS groups. Figure 3.2 and Table 3.2 show that there is no major difference between the mean length of possession between parcels in these two groups, but also their distributions look very much the same.

Moreover, Figure 3.3 shows that the median value of the difference (in years) between the time of possession and the time with a MTS document is located at zero, which means that in most cases the MTS document for the parcel is acquired when the farmer started to work on that piece of land.

As mentioned in the introduction, we also want to know if the probability of having a MTS document on a parcel before the program started is related to some characteristics of the farmers that would indicate a selection process to acquire them. In particular, we explore the possibility that farmers that are better-off are most likely to have gotten a MTS document, so that the informal way of building land rights might be a constrained one. Table 3.3 shows the result of a Probit regression that explains the probability of having a MTS document on

[26] The classification of documents into LTS or MTS could be disputed for the ones located close to the middle of this continuum. Therefore, we create alternative classifications by changing these documents from one group to the other. The results presented in this study were not altered by these changes.

[27] According to the law, possession of these types of documents allows farmer to register its ownership rights in the public system and in fact a small amount of parcels in our sample report having done this by their own but all of them before the PETT Project started. The procedure at that time was consider to be very expensive and time consuming so that probably only farmers with high wealth levels or good connections were able to do it. As these parcels are not part of the PETT program and can introduce some bias in our estimations, we are not considering them in the analysis.

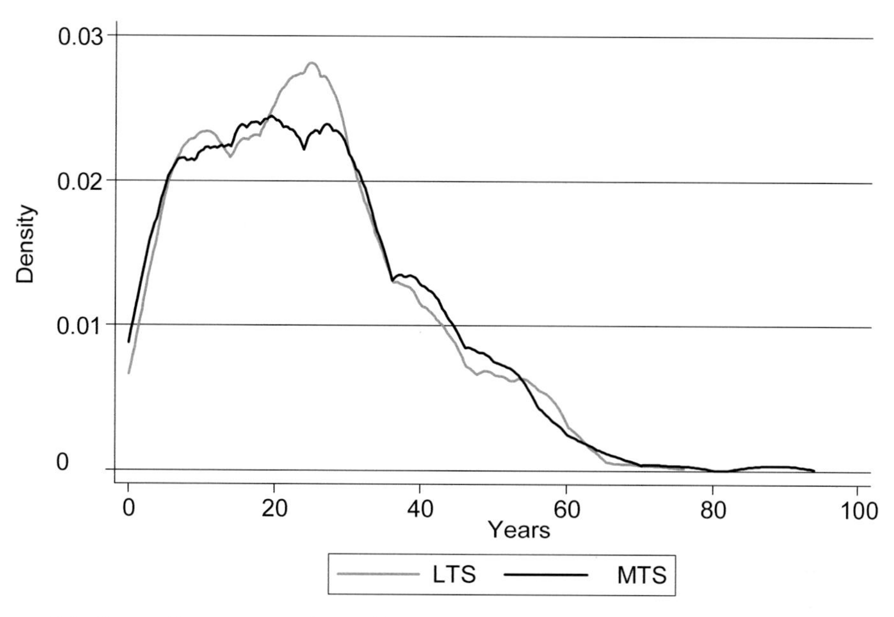

Figure 3.2. Years of possession of parcel by groups.

Table 3.2. Differences in years of possession by groups.

Two-sample t test with equal variances			
	LTS	**MTS**	**t-value**
Years of possession	24.37	24.61	0.6

Two-sample Kolmogorov-Smirnov test		
Smaller group	**D**	**P-value**
0	0.0325	0.049
1	-0.0278	0.11
Combined K-S	0.0325	0.097

a parcel before the start of the program, in terms of some indicators of household's wealth, educational attainment, and market integration.[28]

[28] Total farm size and the number of household members at 1994 are constructed from the survey by using retrospective questions on land transactions and migration, respectively.

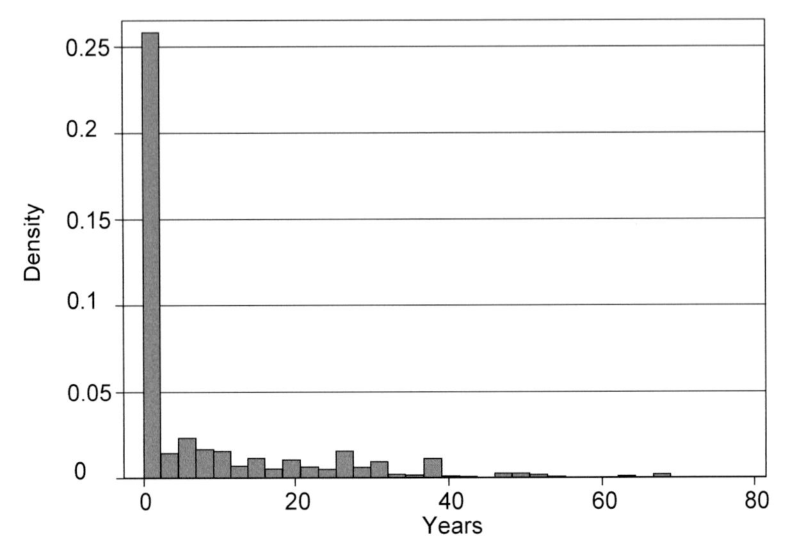

Figure 3.3. Difference between time of possession and time with a document (years).

Table 3.3. Probability of having a MTS document in the parcel at 1994.

Variables	dy/dx	
Total farm size (Has. 1994)	0.020***	(0.008)
Number of household members	-0.004	(0.006)
Sex head of household	0.022	(0.025)
Spanish main lenguage	0.181***	(0.031)
House located in the parcel	0.059*	(0.031)
Time from parcel to district's capital (hours)	-0.065***	(0.007)
Dummy Center-South Coast	0.174***	(0.030)
Dummy North Andean	0.281***	(0.025)
Dummy Center Andean	0.294***	(0.032)
Dummy South Andean	0.296***	(0.027)
Observations	2,300	
Pseudo-R2	0.055	

Robust standard errors between parentheses.
* significant at 10%; ** significant at 5%; *** significant at 1%.

As we can see, the probability of having a MTS document increases with the total size of the farm and with the proximity of the parcel to the capital of the district. Moreover, households where the head has Spanish as their main language, a variable strongly correlated with educational attainment, are also more likely to have acquired one of these documents. Finally, Figure 3.4 shows the percentage of parcels with investments by groups (before T&R) since 1990 until 2000, and confirms the relationship between stronger informal documents and higher initial levels of investments.

By using this classification of parcels we can create the following scenario for testing our hypothesis about the effect of T&R on investments for the two different groups of beneficiaries and controls (see Figure 3.5). A detailed explanation of this figure, its construction using the data from our sample, and the estimator we derive from it, is provided in section 5.

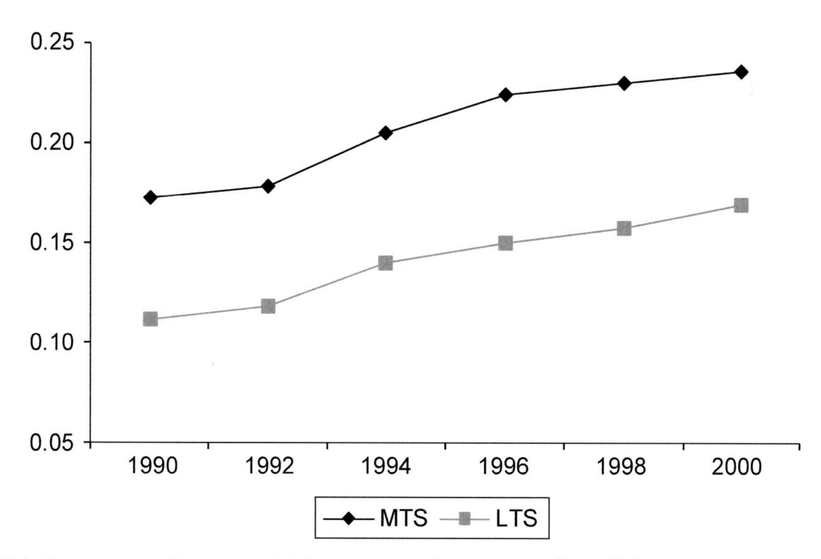

Figure 3.4. Percentage of parcels with investments, by group before T&R.

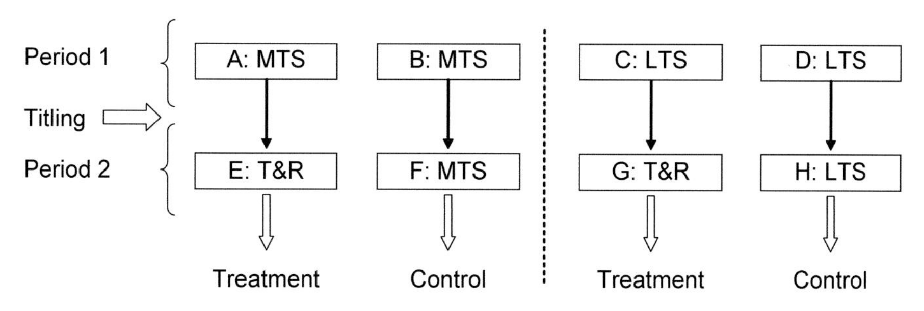

Figure 3.5. Methodological framework.

In the next section we make use of a theoretical model built by Besley (1995) to formalize out hypothesis, and to develop an estimation technique that is consistent with the predictions of the model and the adjustable to the available information.

3.4 Analytical framework

Consider an individual deciding at each period t how much capital, denoted k_t, to invest on a given parcel of land. The returns function for period $t+1$ is $V(k_t, R_{t+1})$, and depends on property rights at $(t+1)$. However, if land rights are exogenously given, and in particular if the decision to invest in one period does not affect future land rights, the returns function can be also written as $V(k_t, R_t)$. Given the context of our study, we will follow this approach.

It is assumed that $V(.,.)$ is increasing in both arguments and concave in k_t.[29] The cost of the investment is denoted by $c(k_t, R_t)$, which is assumed to be increasing in k_t and non-increasing in R_t[30]. The optimal investment choice thus satisfies:

$$\max_{k_t} W(k_t, R_t) \equiv V(k_t, R_t) - c(k_t, R_t) \quad (1)$$

Solving the maximization problem we obtain[31]

$$\frac{\partial k_t}{\partial R_t} = -\frac{W_{12}(k_t, R_t)}{W_{11}(k_t, R_t)} \quad (2)$$

As $W_{11} < 0$ at a maximum, equation (2) implies that an improvement in property rights increases investments if $W_{12} > 0$, that is, if a change in rights increases the marginal return to capital. As we mentioned in the introduction, the two main mechanisms for this condition to be met are the ones related to an increase in farmer's tenure security and the collateral-based credit supply effect[32]. In this chapter we are particularly interested in exploring the 'tenure security argument' given the hypothesis that we want to test. Nevertheless, when analyzing the results, we will make use of some statistical techniques to make sure that the effect is independent of a potential improvement in credit opportunities.[33]

[29] Note that this condition implies decreasing marginal returns of capital.

[30] It is also possible to assume that the cost function is decreasing in R_t so that we incorporate here a simplified version of the collateral-based effect of land rights. Second derivatives of the cost function with respect to both arguments would have to be set equal to cero to maintain the condition that $W_{11}<0$.

[31] Result from the total differentiation of second order.

[32] Besley (1995) explores yet other possibility to meet this condition: the 'gains from trade perspective' asserts that when superior transfer rights lower the cost of exchange if the land is rented-out or sold, improvements made trough investment can be better realized thereby increasing its expected return. The potential importance of this mechanism is explored in detail in Chapter 5.

[33] See Chapter 4 for further analysis on this topic.

The security argument assumes that in period (t+1) there is some chance that an individual has its land expropriated or lost in a legal dispute, and that the probability of this happening is a decreasing function of the rights that he enjoys. This probability can be expressed as a function $\tau(R_t)$ ($\in (0,1)$), where $\tau'(R_t) < 0$. The expected return to investing is $V(k_t, R_t) = [1 - \tau(R_t)] F(k_t)$, where the physical return to the investment is $F(k_t)$, and it is assumed that if the land is lost all the returns are lost with it. By differentiating the returns function with respect to capital first and then with respect to property rights we obtain

$$V_{12} = -\tau'(R_t) F'(k_t) > 0 \tag{3}$$

And under the assumption that the costs are independent of R_t, this result implies that $W_{12} > 0$. Besley (1995) specifically mentioned that this approach may be relevant for the modern Latin America context where squatters who have gained some right to land through prolonged residence face a threat of eviction.

We can think of our particular case as observing a farmer's investment decisions in two consecutive periods of time, with an exogenous change on his property rights in between periods. Parcel's initial rights are heterogeneous in terms of the level of security that they provide, and thus they can be classified into two groups accordingly: low tenure security parcels (LTS), and Medium tenure security parcels (MTS). MTS parcels confront a lower risk of expropriation and hence the marginal return to capital, as well as the propensity to invest, is higher as compared to LTS parcels. Under this scenario, the effect of the new assignment of rights can be derived from the observation on the change in the propensity to invest between period 1 and period 2. If the new rights do not contribute at all to increased tenure security over the parcels, we will expect no change in the propensity to invest on parcels of both groups. If instead, the level of tenure security obtained on MTS parcels was already the maximum possible, we can anticipate no change for this type of parcels while a positive change is expected for parcels in the LTS group. Finally, if the new rights enhance tenure security levels for both groups of parcels, implying that their levels are 'homogenized' at a maximum, we may expect positive changes in both groups but also a higher increase on LTS parcels as these were more constrained before. In the next section we develop an estimation technique to test these hypotheses.

3.5 Econometric model and estimation strategy

The estimator that we want to implement is called the Difference in Difference estimator (DID) and according to Figure 3.5 it would be equal to [(E-A) – (F-B)] for the parcels with initial MTS, and to [(G-C) – (H-D)] for the parcels with initial LTS. The idea is to compare the change in land-attached investments before and after the parcel was T&R with the same change for the relevant control group. To measure this effect, we focus on parcels that were

T&R by the program between years 1994 and 2000[34], and divided them into two groups according to the type of document they hold before titling (LTS or MTS as in the Figure 3.5). The control groups are then conformed by parcels with the same type of document on Period 1 and which have not being titled yet at Period 2.

Retrospective data on land-attached investments include fixed investments in different types of installations, like warehouses, cattle yards, mills, drainage works, water canals, or fences; and land improvements, like terraces or land-grading. Following the predictions of the theoretical model, we expect a farmer to undertake one of these investments if the expected return for doing so is positive. As T&R on a parcel is supposed to increase this return, we will expect a higher proportion of these parcels with investments in period 2 as compared to period 1. Consequently, we measure investments as a discrete variable equal to one if an investment was undertaken in a particular parcel at period *t* and zero if no investment was made. Investments are then recorded for the period 1990-1994 (Period 1) and the period 2000-2004 (Period 2) in order to estimate the difference in the proportion of parcels undertaking land-attached investments before and after the treated parcels in our sample were T&R. As a way of verifying that this change is not biased towards less valuable investments, we also generate a variable for the value of those investments using auto-reported information on the money spend on its construction.[35]

The decision to select as treated parcels only the ones T&R during the period 1994-2000, excluding from the analysis parcels T&R between the years 2001 and 2004, mainly responded to the assumption that land-attached investments of the type used here are not undertaken continuously but rather sporadically. Given the retrospective nature of our data, and in order to have a relatively long period for registering investments, this selection was considered to be optimal.

Because these changes over time may be only reflecting a natural increase in the propensity to invest or renovate investments, or also any other time trend associated with the chosen periods, we will make use of our control group to calculate DID estimates of the effect of T&R on these variables. We are assuming then that the change in the situation of the control group between Period 1 and Period 2 is a good approximation of the change that would have experienced the treated group in this period if they would not have received the title.[36]

[34] This time period covers the whole 1st phase of the PETT Program.

[35] Data on the magnitude of investments, however, might be likely to suffer from problems of measurement errors. The discussion of results and implications will be based only on the findings related to the incidence of land-attached investments.

[36] It is important to notice that the final selected sample consists of a balanced panel of parcels that belong to the same owner in both periods of time. A small amount of parcels (4% of total sample) were acquired after the year 1990 (had only information for period 2) and consequently were excluded from the analysis.

The equation that we estimate is the following[37]:

$$I_{it} = \alpha_0 + \alpha_1(post)_t + \alpha_2(TR)_i + \alpha_3(post*TR)_{it} + \beta' X_{it} + \varepsilon_{it} \qquad (4)$$

The period dummy *post* is equal to zero for all observations on period 1 (1990-1994) and equal to one for period 2 (2000-2004), and it captures any aggregate factors that affect investments over time and in the same way for both the treatment and control groups. Variable *TR* identifies parcels in the treatment and control groups, and X_{it} is a vector of parcels and households characteristics that could be also influencing the decision to invest. As we are interested in the effect of T&R on parcels with a MTS document and parcels with a LTS document, this equation will be run separately for each group. The coefficient on the interaction between *post* and *TR*, α_3, is the estimated program effect, which provides a measure of the conditional average change in investments by treated parcels.

Recent econometric studies suggest that in non-linear models the magnitude of the interaction effect is different from the marginal effect of the interaction term (Norton, 2004), and to compute the real magnitude of the interaction effect one must calculate the cross-derivative of the expected value of the dependent variable. As the sign and values of this interaction effect might be different for different values of the explanatory variables, we report in Appendix 3.1 the sample averages for these parameters. Additionally, we run a linear probability model[38] as an alternative estimation that allows us to verify our findings (Appendix 3.2).

Table 3.4 presents a comparison between the treatment and control groups for a set of parcel and household characteristics that might also influence investment decisions:

A preliminary look at the changes in Investments between periods and groups reveals already an interesting finding. While the percentage of treated parcels with investments increases over time for both groups, control parcels remain unaffected. In terms of other characteristics, the comparison between treated and control parcels for both groups only reveals a few small differences, mostly related to the accessibility of parcels in the treatment groups. For MTS parcels, there seems to be a small but significant difference in the households' head level of education and the percentage of them with Spanish as their main language. The incorporation of these variables in the regression analysis (vector X_{it}) provides a simple way to adjust for observable differences between the different groups, and may also improve the efficiency of the estimate of α_3 by reducing the residual variance.

[37] A detailed explanation of this estimation technique can be found in Meyer (1995) and Wooldridge (2002, pg.129).

[38] This estimation is based on Wooldridge (2002) pg. 454: The Linear Probability Model for Binary Response, who suggests using a weighted least squares regression.

Table 3.4. Summary statistics.[1]

	MTS			LTS		
	Treatment	**Control**	**t-value**	**Treatment**	**Control**	**t-value**
	(1)	**(2)**	**(1,2)**	**(3)**	**(4)**	**(3,4)**
Investments (1990-1994)	0.05	0.07	-1.23	0.03	0.06	-2.11
Investments (2000-2004)	0.10	0.07	2.17	0.10	0.06	1.69
Investments S/. (1990-1994)	6.77	14.09	-1.66	16.15	17.08	-0.08
Investments S/. (2000-2004)	27.81	35.61	-0.38	46.33	20.75	1.23
Parcel characteristics						
Parcel in Altitudes	0.21	0.19	0.76	0.16	0.18	-0.61
Erosion Index (0 no problem, 3 strong erosion)	0.40	0.33	2.06	0.36	0.33	0.81
Slope Index (0 no problem, 2 pronounced)	0.73	0.69	1.09	0.57	0.62	-1.05
Soil Quality Index (1 very bad, 5 very good)	3.23	3.15	2.53	3.18	3.16	0.47
Parcel size (has)	1.38	1.68	-1.21	1.70	1.36	1.44
Time from house to parcel (hours)	29.64	24.96	2.27	29.67	25.84	1.81
Road access to parcel (1 if paved road)	0.08	0.15	-3.94	0.12	0.19	-2.82
Household characteristics						
Hosehold size	3.81	3.91	-0.90	4.20	4.01	1.46
Sex head of household	0.85	0.80	1.16	0.82	0.83	-0.35
Age head of household	63.36	63.28	0.11	60.44	63.79	-3.94
Education head of household (years)	4.28	4.85	-2.66	4.33	4.21	0.48
Spanish main lenguage	0.56	0.66	-3.45	0.56	0.62	-1.72

Note: the value of Investments is reported in Peruvian Nuevos Soles (S/.) at 2004. 1 US$ was equal to 3.4 S/.

[1] It is worth noting that all the variables that are not time-invariant (like sex, location of parcel, or main language) are measured at the time of the survey (2004). To avoid potential endogeneity problems we do not include as controls variables that could have also being affected by the program intervention.

3.6 Results

Table 3.5 presents the DID estimates for the probability of having made a land-attached investment in the parcel (Probit Model) and also for the value of investments (Tobit Model),

Table 3.5. Difference in difference estimation, full sample.

	Probit^				Tobit^^			
	MTS		LTS		MTS		LTS	
Post (α_1)	-0.02	(0.014)	-0.002	(0.014)	5.4	(17.3)	-5.2	(23.8)
TR (α_2)	-0.027*	(0.015)	-0.046**	(0.019)	-34.8*	(19.7)	-64.8**	(30.4)
Post*TR (α_3)	0.054**	(0.027)	0.085***	(0.036)	54.1**	(29.1)	107.6***	(45.1)
Observations	2,186		1,547		2,186		1,547	

Standard errors reported between parentheses.
* significant at 10%; ** significant at 5%; *** significant at 1%.
^ Marginal effects from probit estimates reported.
^^ Conditional marginal effects of tobit estimates reported.

controlling for the set of parcel and household characteristics presented before, plus dummy variables for the regional domains.

As we can see, there is a positive and significant effect of T&R (α_3) on the propensity to invest as well as on the value of investments for both the MTS and LTS group of parcels. The coefficients on the MTS group, however, are smaller in both regressions than the ones for the LTS group. In terms of their pre-program situation, receiving a title from the program is associated with doubling the probability of making an investment for MTS parcels, while the effect of titling augmented in more than four times the probability of making investments on LTS parcels. This result confirms the presence of a 'homogenization effect' of T&R on investments.

Appendix 3.1 presents the results for the (average) interaction effect calculated using the Norton *et al.* (2004) method. Even though the estimates of the interaction effect for both groups are reduced, they are still significant on average. Moreover, the results of the linear probability model in Appendix 3.2 show only a slight difference in terms of the magnitude of the α_3 coefficient from the DID regression[39].

In order to verify whether these results reflect a change in the tenure security perception of farmers, or if they can be attributable to an improvement in credit access brought by titling,

[39] The difference between the number of observations used in this regression and in the DID regression is due to the method applied to construct variance weights for the linear probability model. Approximately 5% of the total sample of parcels had to be excluded for having a negative predicted value of the dependent variable. The comparison between a simple OLS regression with and without these observations yields almost no difference in the value and significance of parameters.

we purse two alternative strategies.[40] First, we distinguish investments financed with credit from those financed out of pocket (OOP), to test the effect of titling on the latter category alone. This distinction is based on the respondent's answer to the questions about the type of financing used for each investment that was made on a parcel. This strategy, however, does not rule out the possibility that improvements in credit access could have been used for other purposes, thus affecting investment incentives in an indirect way. However, as 97 percent of the reported investments were said to be financed OOP, we do not expect a significant difference when following this approach. Second, a related test can be performed by isolating the effect of titling on investments amongst non-borrowing households. Only a 7 percent of the households in the sample reported having access to formal sources of credit over the past three years. Even though our survey did not record access to credit prior to the program intervention, including borrowing households at that time does not invalidate our test as these households would not show an increase in credit access after the program. If the investment effect is robust to this limitation, we can confirm that the principal mechanism at work is the one related to a change in tenure security.

As we can see in Table 3.6, the estimates for the OOP investments are almost identical in magnitude and significance as the ones obtained when the total number of investments was used. Limiting the sample to non-borrowing households increases the α_3 coefficient by one percent for the MTS parcels, while reduces it by the same amount for the LTS parcels, but these changes are not statistically significant when compared to the α_3 coefficients in the full sample of parcels.[41] These results suggest that the increase in investments is almost entirely driven by higher levels of tenure security brought by the title.

3.7 Concluding remarks

The results of this chapter indicate that land titling policies aiming to formalize individual land rights have a differentiated effect on investments, depending on the farmer's level of tenure security over a parcel before the policy was initiated. We show that before the intervention of the program, parcels can be already categorized into different levels of tenure security depending on the type of informal documents that farmers hold. Accordingly, parcels with 'stronger' documents present initially higher levels of investments compared to parcels with 'weaker' documents. The effect of the titling policy on the propensity to invest and on the value of investments is positive and significant for both groups, but shows a stronger impact

[40] The study by Field (2005) applies a similar test to distinguish changes in ability versus changes in the willingness to invest for households participating in an Urban Land Titling Program. Besley (1995) suggest that the collateral effects can be distinguish by adding a dummy variable equal to one if the household has at least one parcel titled. As the PETT program titled at the same time all parcels located in the same valley, the households in our sample have either all or none of their parcels titled during the period 1994-2000, so that this method cannot be implemented here.

[41] The test for equality of the coefficients across samples reports a chi2(1) of 1.59 and a corresponding p-value of 0.21 for the MTS parcels, and chi2(1) of 0.23 and p-value of 0.63 for the LTS parcels.

Table 3.6. Difference in difference estimation, OOP investments and non-borrowers.

	Probit				**Tobit**			
	MTS		**LTS**		**MTS**		**LTS**	
OOP Investments								
Post (α_1)	-0.001	(0.013)	-0.004	(0.014)	6.4	(17.4)	-9.6	(23.8)
TR (α_2)	-0.028*	(0.015)	-0.044**	(0.018)	-36.3*	(19.9)	-61.8**	(30.2)
Post*TR (α_3)	0.056**	(0.028)	0.088***	(0.036)	56.8**	(29.5)	111.9***	(45.2)
Observations	2,186		1,547		2,186		1,547	
Non-borrowers								
Post (α_1)	-0.005	(0.014)	-0.003	(0.014)	2.2	(17.9)	-7.8	(24.8)
TR (α_2)	-0.033**	(0.016)	-0.048***	(0.019)	-40.6**	(20.7)	-69.8**	(32.3)
Post*TR (α_3)	0.064**	(0.03)	0.076***	(0.036)	62.5**	(31.0)	104.1**	(47.8)
Observations	2,039		1,437		2,039		1,437	

Standard errors reported between parentheses.
* significant at 10%; ** significant at 5%; *** significant at 1%.

on parcels with previously weaker levels of tenure security. Moreover, this effect can be almost entirely attributed to changes in farmer's willingness to invest and not to better access of credit.

We expect that these results contribute to the debate about the need for a public intervention in the formalization of land property rights, particularly in Latin America. Even though farmers can get access to informal land documents to increase their security over the land, we show that this procedure is mostly limited to farmers that were already better-off and it constitutes at best an imperfect substitution to the acquisition of full-fledged property titles like the ones provided by the PETT program. The differentiated effects of the title on investments between MTS and LTS parcels, reinforces this idea and argues in favor of the importance of a public intervention like this one to lift-up the limitations for disadvantaged farmers to acquire tenure security by informal means. The recognition of different sorts of informal land rights and the reliance of the program on community networks before the formalization of rights also appear to be fundamental for a successful intervention with a promising pro-poor orientation.

Finally, it is important to notice that even though we found a justification for this type of intervention, there are many other aspects of the titling policy that need to be analyzed in order to fully assess its potential effects and limitations. For example, the fact that the new investments brought by titling were mostly financed without the use of credit could indicate that they are limited to small, and probably labor-intensive activities, which might not have a large impact on factor productivity or land values. When farmers were asked about their willingness to pursue more investments in land, and their principal constraints to do so, many of them pointed to the lack of credit as the main reason. Therefore, more work needs to be done in order to explore the constraints that farmer's face in other markets which can be influencing the potential effects of the program (see Chapter 4).

Appendix 3.1. Interaction effects of probit model

	MTS			
	Mean	**Std. dev.**	**Min**	**Max**
Interaction effect	0.048	0.023	0.004	0.127
Standard error	0.024	0.010	0.003	0.058
z-value	1.936	0.203	1.044	2.251
	LTS			
	Mean	**Std. dev.**	**Min**	**Max**
Interaction effect	0.062	0.027	0.013	0.221
Standard error	0.029	0.011	0.009	0.093
z-value	2.106	0.289	1.309	2.688

Appendix 3.2. Linear probability model

	MTS		LTS	
Post (α_1)	0.000	(0.017)	-0.004	(0.018)
TR (α_2)	-0.032*	(0.019)	-0.073***	(0.024)
Post*TR (α_3)	0.057**	(0.025)	0.091***	(0.029)
Observations	2,061		1,476	

Standard errors reported between parentheses
* significant at 10%; ** significant at 5%; *** significant at 1%

4. Credit constraints in the Peruvian rural sector: can titling provide a solution?[42]

Abstract

This chapter explores the characteristics of supply and demand for formal loans in the Peruvian agricultural sector, and analyzes the principal determinants and constraints that farmers face for accessing this source of credit. Special attention is placed on the potential effect of the Peruvian Titling Program on lifting up some of these impediments and increasing credit access for its beneficiaries. We use survey questions specifically designed to identify rationing mechanisms for each individual, and a multinomial logit regression to determine the probability of being subject to each type of them. Our results show that more than half of our sample of farmers has a positive loan demand that is unsatisfied because of the presence of information asymmetries. While having a registered land title appears to decrease the transaction cost involved in formal loan applications, it is far from being a sufficient condition to get access to a loan in this sector. The existence of multiple limitations from the supply and demand side implies that getting access to a formal loan becomes almost exclusively an option for wealthier farmers, with large amounts of land, and high levels of education. Land titling could facilitate their access and probably improve the conditions of their borrowing contracts, but it does not affect the possibilities for small-scale and poor producers.

4.1 Introduction

The provision and registration of land titles has been hypothesized to have a direct impact on farmer's access to credit because of its effect on increasing the collateral value of land for credit lenders. This effect will be especially true regarding formal credit sources which often have imperfect information on borrowers and thus insist on collateral before advancing a loan. However, a large amount of evidence suggest a weak or even null impact of titling programs on credit access, particularly in Latin America (Boucher *et al.*, 2004; Guirkinger and Boucher, 2006). In addition, in the few cases where a positive effect could indeed be established it was found to be mostly in favor of wealthier producers (Aldana and Fort, 2001; Carter and Olinto, 2003).

As Platteau (2000) mentions, low credit use may actually be caused by two distinct types of factors. On the one hand, it may result from supply failures that have their origin in various imperfections, not only in the credit market itself but also in other rural factor markets, particularly in the land market. On the other hand, it may be determined by demand failures that prevent farmers from tapping available credit sources.

[42] Part of this study was presented at the International Conference on Land, Poverty, Social Justice, and Development in the Institute of Social Studies, The Hague, January 2006.

Failure to provide credit in spite of titling may arise, for example, if titled land is not considered a reliable collateral by credit lenders because it is difficult to foreclose or when it is difficult to dispose of the land in case of default if local land markets are thin (Okoth-Ogendo, 1976; Collier, 1983; Noronha, 1985; Bruce, 1986; Barrows and Roth, 1989). Besides these difficulties, there might be other supply-constraints arising from the strategies of credit providers. First, commercial banks and financial institutions are often reluctant to lend for land purchase because they are unwilling to tie up their capital, raised largely trough short-term deposits, for long periods of time. Moreover, bankers usually prefer lending against more reliable streams of income than those found in agriculture. Second, considerations of administrative costs may lead banks to set a minimum size of loans which often exceeds the capital needs of small farmers, or to refuse to lend to them on ground that their property is costly to dispose of in the event of foreclosure due to the tiny size of fragmented landholdings.

From the demand side, farmers may also fail to apply for loans due to different reasons. High transaction costs involved in loan applications, as well as lack of information or high costs in the application process could prevent farmers from applying (Chung, 1995; Mushinski, 1999). These reasons will probably be more important for farmers that require smaller amounts of credit, have lower levels of education, and are distant from markets. Farmers may also fail to apply for loans because they perceive a high risk of losing their land through foreclosure. Boucher and Carter (2002) have labeled this option as 'risk rationing'. As they explain, this outcome occurs when lenders, constrained by asymmetric information, shift so much contractual risk to the borrower that he voluntarily withdraws from the credit market even when having the necessary collateral wealth to qualify for an incentive compatible loan contract. Also, under mild assumptions about the nature of farmer's risk aversion, risk rationing will be wealth-biased and predominately affects lower wealth individuals. Finally, similarly to farmers that do apply for these loans and are rejected for not having enough collateral, many others may restrict themselves from applying because they believe there is a strong chance of rejection. This mechanism has been labeled as 'quantity rationing' in the specialized literature (Stiglitz and Weiss, 1981; Carter, 1988), and tends to be stronger for farmers without clearly defined property rights over their land, or farmers with small land-holdings.

The purpose of this chapter is first, to explore the characteristics of supply and demand for formal loans in the Peruvian agricultural sector, and second, to analyze the principal determinants and constraints that farmers face for accessing these sources of credit.[43] Special attention is placed on the potential effect of the Peruvian Titling Program for lifting up some of these impediments and increasing credit access for its beneficiaries. Based on these results we expect to be able to identify complementary policies required for improving farmers, and particularly small-farmers, access to financial resources.

[43] Previous studies showed that these rationing mechanisms operate in the formal sector but not in the informal one Moreover, formal loans in Peru usually provide higher amounts, longer credit terms, and lower interest rates than loans in the informal sector (Trivelli and Venero 1999; Guirkinger 2005).

The rest of the chapter is organized as follows. Section 2 provides a summary of the main characteristics of formal credit supply in Peru, and explores its fundamental limitations. Next, we turn the analysis to the farmer's perspective, and discuss a methodology that allows us to classify them into different credit rationing categories using information from our household's survey. After an initial exploration of the magnitude of these rationing mechanisms, and its relationship with some household's characteristics, we explain the estimation technique used to analyze their determinants. Estimation results are presented in section 4; and section 5 gives the concluding remarks.

4.2 Supply of formal credit in Peru

The economic liberalization process that occurred in Peru at the beginning of the nineties eliminated the participation of the state in the financial system. As a consequence, the Agrarian Bank was forced to close in 1992, leaving unattended a big mass of producers in rural areas. The yearly supply of loans from this institution was around US$ 500 million distributed amongst approximately 250,000 clients.[44] As we can see in Figure 4.1, the amount of loans to the agricultural sector increased in a continuous way until 1998 as a result of the appearance of new financial institutions as the 'Cajas Rurales' (CRAC), the consolidation of the 'Cajas Municipales' (CMAC) that started to work in rural areas, and most of all for the increase in credit supply from banks to large agro-export producers. The biggest difference with the situation during the eighties has to do with the number of clients at the end of the nineties, which are estimated to be around 10% of the number attended by the Agrarian Bank[45].

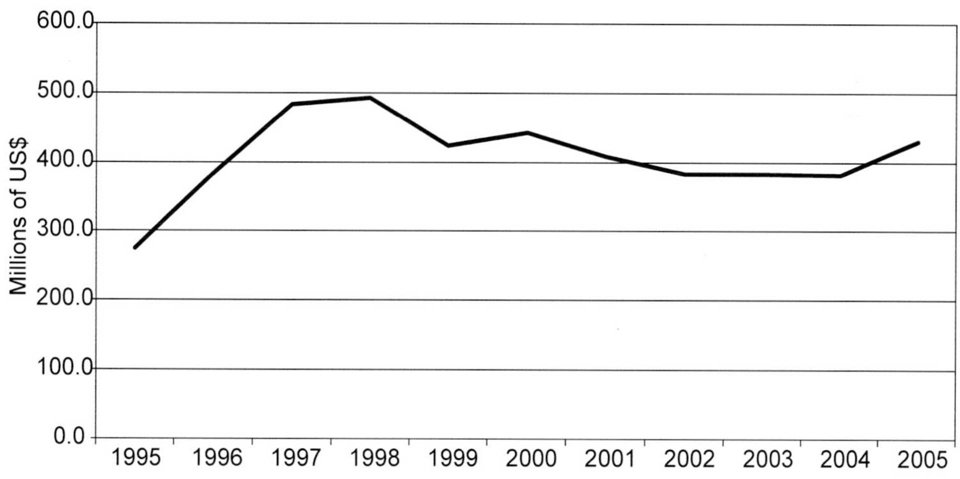

Figure 4.1. Agricultural credit supply from financial institutions (1995-2005) (Source: SBS).

[44] Estimations made by Guirkinger and Trivelli (2005). Amount in current US$.

[45] Guirkinger and Trivelli (2005) estimate in more than 40,000 the number of clients in the formal credit sector at 2004. This big increase appears to be mostly due to the expansion of the CMAC.

The consolidation of the CMAC and CRAC, together with the more recent 'Entidades de Desarrollo para la Pequeña y Microempresa' (EDPYME) has created a new kind of financial institution in rural areas of Peru which scale of operations and credit strategies differ from the traditional banking system. Their principal target was the segment of medium and small-scale producers who were normally not attended by commercial banks or the former Agrarian Bank. Their presence marked the beginning of the micro-credit business in the rural financial market of Peru. Despite this expansion, more than 80% of the loans supply comes from the formal banking system (Table 4.1).

According to the public organization in charge of supervising financial institutions in Peru (*Super Intendencia de Banca y Seguros - SBS*), more than three quarters of the credit from these institutions is concentrated in 6 departments (Arequipa, Piura, La Libertad, Cajamarca, Cuzco, Lima, Lambayeque, Ica, and Junin). Coincidently, these departments are the ones with

Table 4.1. Agricultural credit supply from financial institutions 1995-2005 (Thousands of US$).

	1995	**1996**	**1997**	**1998**	**1999**	**2000**	**2001**	**2002**	**2003**	**2004**	**2005**
Comercial banks	257,135	350,143	436,771	438,219	373,397	391,363	350,277	331,967	330,542	320,192	362,057
Financial institutions	2,898	1,702	3,425	4,778	1,015	739	1,966	2,620	969	3	0
CMAC	3,152	6,387	8,769	11892	14,898	15,015	17,698	16,758	18,338	24,144	28,101
CRAC	10,964	23,620	32,715	36,529	33,892	34,360	35,527	29,765	28,736	28,903	31,120
EDPYMES							1,501	1,199	3,187	6,560	7,062
TOTAL	274,149	381,852	481,679	491,418	423,203	441,477	406,968	382,307	381,772	379,802	428,339

Source: Aguilar (2003) and SBS since 2001

Table 4.2. Agricultural credit as a percentage of total credit supply by institution (%).

	1995	**1996**	**1997**	**1998**	**1999**	**2000**	**2001**	**2002**	**2003**	**2004**	**2005**
Comercial banks	3.3	3.4	3.3	3.1	3.1	3.3	4.0	3.9	4.3	4.0	4.1
Financial institutions	1.1	1.4	1.6	1.5	0.6	0.4	1.8	2.4	1.1	0.1	0.0
CMAC	5.5	8.5	9.2	11.5	11.9	8.6	11.1	7.4	5.7	5.4	5.1
CRAC	47.1	60.1	59.3	63.5	61.7	54.4	66.0	51.7	43.8	33.5	31.3
EDPYME			10.2	4.2	2.4	4	3.4	2.2	4.3	6.6	5.5
TOTAL	3.4	3.6	3.5	3.4	3.5	3.6	4.4	4.2	4.6	4.4	4.5

Source: Aguilar (2003) and SBS since 2001

higher titling density levels[46] and also where land markets presented a more dynamic pattern in the last decade.[47]

To further explore the potential relationship between tilting and credit opportunities, we present information from our sampled Districts (70) and combine this with credit supply data from the SBS. We use the percentage of parcels that are formally titled and registered in each district (titling density) and compare it to two measures of credit supply available at the same level. One of them is the number of formal credit institutions (Banks, CRAC, CMAC, and EDPYMES) located in the Province where the District belongs[48], and the other is the travel time from the capital of the District to the location of the closest formal credit institution's office. This last measure was constructed using geographical coordinates for the District's capital center and for the location of the credit offices, and then simulating the time of traveling by car between them based on complementary information about the altitude, steepness, and type of road.

The results presented in Figures 4.2 and 4.3 show that there is a positive correlation between the level of titling density on a District and the number of formal financial offices located in

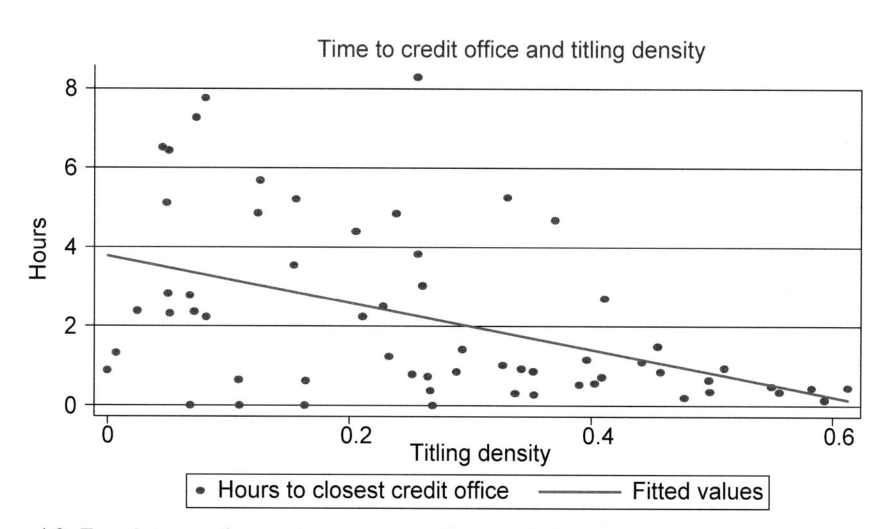

Figure 4.2. Travel time to financial institution's offices and tilting density.

[46] Titling density over a region is understood as the percentage of Titled and Registered parcels over the total number of parcels in that region.

[47] Many of these Departments, particularly the ones located in the Coast, have experienced an agro-export 'boom' in the last decade. Escobal (1998) explores some of the factors that made this possible, and provides information on investments made by valley.

[48] The geopolitical division of Peru has Departments as the highest organization level. These are divided into Provinces, and Provinces into Districts.

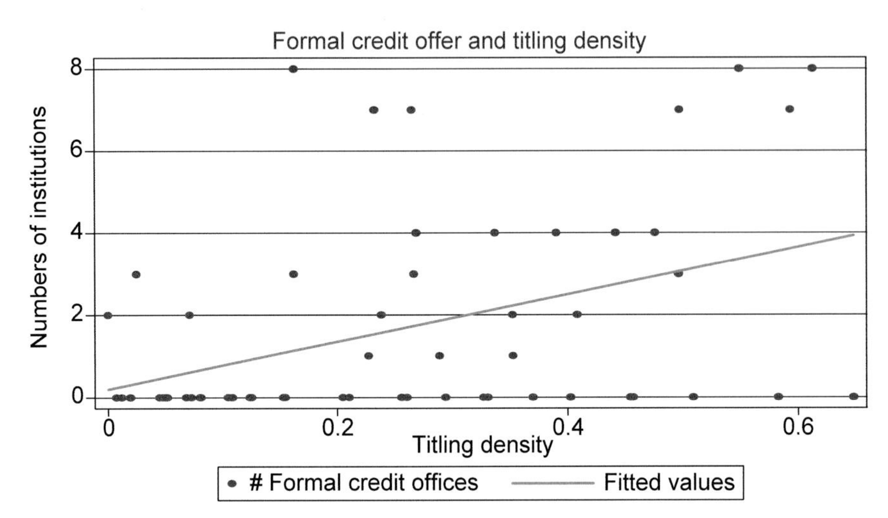

Figure 4.3. Formal credit offices and titling density.

its Province (r=0.38), and a negative one with the time from the Districts' capital to the closest office (r=-0.48)[49].

As we saw in the previous tables, the total amount of credit has not increased in the last years despite the large effort of the Titling Program. This lack of increase in credit supply does not seem to be derived from insufficient funds of formal institutions because SBS reports increasing deposits captured in the last years. The problem seems to be that these institutions prefer not to give loans for agricultural purposes, and that many farmers in this sector prefer not to ask for loans from these institutions.

As Portocarrero and Tarazona (2003) notice in their analysis of two CRAC, 43 percent of their portfolio is concentrated in 8 percent of their largest clients. The main reason for not incrementing their supply of loans, and moreover, to concentrate it on only a few clients is related to the perception of an increased risk and uncertainty associated with the financial policy of the government during the last years. In particular, the approval of the Agrarian Financial Rescue Program ('Programa de Rescate Financiero Agrario'-RFA) and the creation of a new State Bank for Agricultural Lending (AGROBANCO) are perceived as negative interventions that increase the probability of a client defaulting on a loan, or deciding to refinance unilaterally their obligations[50]. Therefore, some credit institutions argue that they prefer to restrict their loans to 'healthy clients' who have repaid their loans in spite of having confronted negative shocks, and of having the option to use the RFA mechanisms.

[49] Values of **r** represent the Pearson's correlation coefficient.

[50] After the negative events produced by 'El Ninho Phenomenon' of 1998, the government create the RFA whit the intention to refinance agricultural debts with the use of government bonds. As Guirkinger and Trivelli (2005) mention, the rate of defaults in the formal credit sector augmented with this measure.

In addition, micro-finance institutions seem to be reducing their exposure to the agricultural sector. As some studies suggest (Portocarrero and Tarazona, 2003; Guirkinger, 2005), this fact is related to new diversification strategies that present more advantages for these institutions, and particularly for the CRAC. By evaluating their portfolio costs in each sector, they can find higher profitability and lower risk for loans allocated in urban areas and business as compared to the ones in the agricultural sector. Even in rural areas, they seem to have found better margins when allocating loans to non-agricultural business, which has contributed to the further reduction in supply credit for the sector.

4.3 Credit constraints from farmer's perspective

The asymmetric information problem that is endemic to credit transactions gives rise to the potential for non-price rationing. This makes empirical analysis challenging and conventional techniques non applicable to credit markets because the observed loan quantities do not necessarily represent the intersection of supply and demand. As such, researchers may not be able to infer the rationing mechanism at work just by observing the transactions. When prices cannot freely adjust to clear a market, the information provided by observing a transaction implies that the quantity transacted represents the minimum between supply and demand.[51] Even though the observation of a positive loan amount can provide some information on demand and supply conditions of a particular individual, the case of non-borrowers turns to be more problematic. Theoretical models such as those by Stiglitz and Weiss (1981) and Carter (1988) explain how non-price rationing may also take the form of complete rejection such that the available supply for an individual is zero. These 'quantity rationed' borrowers have a positive demand for loans at the current contract conditions, and thus need to be differentiated from price rationed individuals who have zero demand. One approach[52] followed recently by many scholars is to design their survey as to directly collect this information from qualitative questions designed to identify rationing mechanisms for each individual.[53] We relied on this methodology and included several qualitative questions in our survey in order to recover information on different sources of credit rationing. The classification of farmers by rationing categories follows the method employed by Boucher (2002) in his study of the North-Coast of Peru, and its adaptation to our sample is explained here in detail (see Table 4.3).

Farmers that applied for a loan were classified into two different categories. Those that received the loan were grouped as price rationed with loan (prwl), and those who had their application rejected were classified as quantity rationed (qrat). Non-applicants were asked the following hypothetical question: 'If you would have applied to a commercial lender, would the lender

[51] This idea has being derived from the econometric literature on disequilibrium models. For its application to credit markets see Feder *et al.* (1988).

[52] The empirical literature presents other mechanisms to deal with this problem that are mostly based on strong assumptions on how credit markets work in rural areas. For a review of these approaches see Boucher (2002).

[53] This approach has been used in Feder *et al.* (1990), Barham *et al.* (1996), Mushinsky (1996), to mention some examples.

Table 4.3. Description of rationing categories.

Rationing category	Mechanism description
1. Price rationed with loan (prwl)	Applied for a loan and received it
2. Price rationed without a loan (prwol)	Did not apply for a loan because interest rate was too high
3. Quantity rationed (qrat)	Applied for a loan and was rejected or did not apply because subjective probability of rejection was too high
4. Risk rationed (rrat)	Did not apply for a loan for fear of losing the collateral
5. Transaction Cost rationed (tcrat)	Did not apply for a loan because transaction costs were too high

have approved the application?' Farmers with a positive answer were then asked about their reasons for not applying and based on their open-ended responses they were classified into price rationed without a loan (prwol), risk rationed (rrat), and transaction costs rationed (tcrat). For example, non-applicants who did not apply because interest rates were too high would be placed as price rationed without a loan (prowl). If the reason for not applying was related to their fear of loosing the land, farmers would be categorized as risk rationed (rrat), and if it was because of the cost of application, or the time and expenses of the application process, they would be placed as transaction cost rationed (tcrat). Farmers who felt that a commercial lender would not have approved their loan, were asked for the reasons to belief so and then classified as quantity rationed (qrat) if they would have liked to have a loan but lack of enough collateral, or into categories 2, 4 or 5 accordingly to their open-ended responses as before.

4.3.1 Rationing mechanisms in the sample under study

We make use of this classification of farmers into different rationing categories to explore their underlying determinants. Our data contains information for 847 rural households located in five different geographic domains of the Coastal and Andean regions in Peru that were interviewed in 2004 as part of the Titling Program's evaluation. About half of them received a registered title from the program on its first stage (1994-2000), while the rest are future beneficiaries that have not yet received their title. The survey' section on credit contains information about loan applications in the last three years, as well as all the perception questions needed to construct the rationing categories.

As discussed before, household wealth and the value of collateral are two important factors to understand farmer's constraints on the credit market. As the most important collateral accepted by formal lenders in rural areas is owned land, we want to analyze the importance of the different rationing mechanisms for farmers with different farms sizes and titling status. We also use the value of all durable goods owned by the household as a measure of household wealth and a signal of repayment capacity for lenders.[54] We make use of retrospective information collected in the survey to construct these variables at the year 2000, trying to avoid reverse causality problems.

The last row of Table 4.4 presents the frequency of each rationing mechanism for all households in our sample, also dividing households into No-Pett and Pett categories. Only 9% of the households in the sample report having received a loan from a formal institution in the last three years (prwl), and together with the ones that didn't apply because the interest rate was too high (prwol) they add up to 40 percent of all households. This means that for around 60 percent of the sample there are other mechanisms at work (non-price rationing) that prevent them from getting a desired loan. Also, the traditional quantity rationing mechanism by which households are rejected or fail to apply because of collateral issues is present for 36 percent of the sampled households, implying that high transaction costs and the risk of loosing the collateral represent also important constraints.

The proportion of households that are rationed by price increases with wealth, as we can notice when moving from the first to the fifth quintile. The percentage of households that are price-rationed with a loan (prwl) goes from 2 percent on the first quintile to 21 percent in the fifth, and from 22 percent to 38 percent for price-rationed without a loan (prwol). While there is no apparent difference between titled and non-titled households in the prowl category, having a title seems to increase the probability of getting a formal loan (prwl) for households in the last two quintiles of wealth.

Table 4.4. Rationing mechanisms by wealth quintiles.

	prwl			**prwol**			**qrat**			**tcrat**			**rrat**		
	All	**No-Pett**	**Pett**	**All**	**No-Pett**	**Pett**	**All**	**No-Pett**	**Pett**	**All**	**No-Pett**	**Pett**	**All**	**No-Pett**	**Pett**
1	2%	2%	2%	22%	25%	18%	44%	40%	48%	13%	18%	9%	17%	13%	20%
2	2%	2%	1%	27%	28%	25%	41%	33%	51%	13%	18%	7%	15%	15%	15%
3	8%	8%	8%	33%	33%	33%	45%	41%	48%	6%	5%	7%	8%	12%	4%
4	12%	8%	15%	39%	38%	41%	28%	31%	25%	9%	14%	5%	9%	8%	11%
5	21%	14%	26%	38%	39%	37%	24%	25%	23%	7%	8%	7%	8%	9%	7%
Total	9%	7%	11%	32%	32%	31%	36%	34%	38%	10%	13%	7%	11%	11%	11%

[54] Our survey recovers information for household's goods as radios, televisions, fridges, kitchens, motorcycles, cars, and many others, together with the year of acquisition, the price at that year, and the price at which they will sell it now.

Contrary to what we observed in price rationing mechanisms, the percentage of households in the quantity rationing category decreases with wealth. While 44 percent of households in the first quintile are quantity rationed, only 24 percent belong to this category in the last quintile. The same pattern is observed for households labeled as transaction cost rationed and risk rationed. Moreover, while having a title does not seem to affect the probability of a household being quantity or risk rationed, it does seem to make a difference in terms of being rationed because of high transaction costs. Larson *et al.* (2003), Boucher (2000), and Trivelli and Venero (1999) estimate that the cost of the proceedings related to the use of land as collateral vary between 5 and 10 percent of the amount of the loan for small producers. This percentage is relatively high considering the average loan for these farmers is small. As the Titling Program incorporates also some components to reduce the administrative process to register land as collateral, these finding might confirm that these costs have actually being reduced.

Basically the same results discussed above hold when dividing the sample of households by quintiles of farm size (Table 4.5). But when the division is made by quartiles of titling density at the District level (Table 4.6), some new relationships are revealed. In this case, the probability

Table 4.5. Rationing mechanisms by farm size quintiles.

	prwl			prwol			qrat			tcrat			rrat		
	All	No-Pett	Pett	All	No-Pett	Pett	All	No-Pett	Pett	All	No-Pett	Pett	All	No-Pett	Pett
1	2%	1%	5%	25%	26%	22%	41%	35%	51%	15%	19%	10%	15%	16%	13%
2	7%	5%	8%	31%	36%	27%	45%	45%	45%	7%	8%	7%	9%	4%	12%
3	9%	8%	9%	33%	29%	36%	34%	35%	33%	13%	14%	11%	11%	11%	11%
4	12%	10%	14%	32%	35%	29%	28%	20%	34%	9%	13%	6%	17%	19%	15%
5	15%	8%	19%	37%	35%	38%	34%	38%	32%	5%	8%	3%	6%	6%	7%

Table 4.6. Rationing mechanisms by titling density.

	prwl			prwol			qrat			tcrat			rrat		
	All	No-Pett	Pett	All	No-Pett	Pett	All	No-Pett	Pett	All	No-Pett	Pett	All	No-Pett	Pett
1	6%	4%	7%	30%	28%	30%	35%	32%	37%	15%	21%	10%	13%	14%	13%
2	7%	6%	9%	22%	20%	25%	43%	42%	43%	10%	11%	8%	14%	14%	14%
3	10%	6%	12%	35%	40%	31%	35%	30%	39%	7%	11%	5%	12%	12%	12%
4	13%	9%	17%	39%	40%	38%	33%	33%	33%	7%	9%	5%	6%	7%	6%

of getting a loan (**prwl**) increases with the level of titling density but only for titled farmers, and also the percentage of farmers in the transaction costs-rationing category decreases when the level of land rights formalization increases. These results are in line with the evidence presented above, where the levels of titling density appeared to be positively correlated with the number of offices from formal financial institutions, and negatively correlated with the traveling time to the closest office. But also, as discussed in Chapter 5 the level of titling density could be affecting farmer's demand for land related investments and therefore their demand for credit. If land markets expand as a result of increasing the general level of tenure security in the region, land becomes a more 'liquid' asset and hence any improvement made trough investments can be better realized in the case that land is transacted.

In order to better understand the importance of these different variables to explain the rationing mechanism at work, we will make use of a multi-variate analysis.

4.3.2 Multinomial logit estimation framework

We use a multinomial choice model to predict the probability that farmers with different characteristics are found in the rationing categories discussed above. The multinomial logit model is considered particularly convenient. In this model, the categorical variable Y represents the observed credit market rationing outcome taking values 0, 1, ..., J. Define Y^*_{ij} as the continuous 'score' for the i'th individual in the j'th rationing category, or the unobserved 'propensity' of the individual i to be in the category j. Modeling it as a linear combination of household's characteristics we obtain:

$$Y^*_{ij} = g(\beta_j, x_i) + \varepsilon_{ij} = \beta_j' x_i + \varepsilon_{ij} \tag{1}$$

where x_i is a (1 x k) vector of characteristics of the i'th individual; β_j is a (k x 1) vector of population parameters associated with the j'th category to be estimated; and ε_{ij} is the unobserved component of the I'th individual's score from category j. The observed category is the one which yield the highest score. The probability that the i'th individual is in the j'th category is then:

$$\Pr(Y_i = j) = \Pr(Y^*_{ij} \geq Y^*_{is})\ \forall s \neq j \tag{2}$$

If the (J+1) ε_{ij} terms are independent and identically distributed with Weibull distribution then the probability in equation (2) can be expressed as:

$$\Pr(Y_i = j) = \frac{e^{\beta_j' xi}}{\sum_{s=0}^{J} e^{\beta_s' x_i}} \tag{3}$$

This multinomial framework allow us to impose a probability structure on the outcomes, and the logistic form is specially suitable because it can capture non-linear relationships between the independent variables and the regime probabilities, and it also keeps the probability within the unit interval. In order to achieve identification of all parameters in the model[55], the values of the parameters associated with one of the categories need to be fixed. The most convenient normalization is to choose a 'base' category and set its parameters equal to cero so that $\beta_0=0$ when j=0. Following that normalization the probabilities in equation (3) become:

$$\Pr(Y_i = 0) = \frac{1}{1 + \sum_{s=1}^{J} e^{\beta_s' x_i}} \tag{4}$$

$$\Pr(Y_i = j) = \frac{e^{\beta_j' x_i}}{1 + \sum_{s=1}^{J} e^{\beta_s' x_i}} \quad , j = 1,2,\ldots,J \tag{5}$$

The parameters of the vector βj in the multinomial regression represent the impact of individual characteristics on the probability of an individual being observed in category j relative to the base category. Using equation (4) and (5) we can see that this implies:

$$\ln\left[\frac{\Pr(Y_i = j)}{\Pr(Y_i = 0)}\right] = \ln\,[e^{\beta_j' x_i}] = \beta_j' x_i \tag{6}$$

So that the elements of βj give the marginal impacts of individual characteristics on the log of the odds ratio, which is just the ratio of two probabilities as seeing in (6). Because our main concern is to explore the effect of titling on participation on the credit market for farmers with different characteristics, our interest does not lie in the coefficient estimates themselves, but rather in $\partial P_{ij} / \partial X$ - the marginal impact of the regressor on each rationing category, where P_{ij} denotes $\Pr(Y_{ij}=1)$. Equation (7) and (8) give the expression for the marginal effects on the base category and j'th category respectively:

$$\frac{\partial P_{0i}}{\partial x} = P_{0i} \sum_{s=1}^{J} P_{si} \frac{\partial g(\beta_s, x_i)}{\partial x} \tag{7}$$

$$\frac{\partial P_{ji}}{\partial x} = P_{ji}\left[\frac{\partial g(\beta_j, x_i)}{\partial x} - \sum_{s=1}^{J} P_{si} \frac{\partial g(\beta_s, x_i)}{\partial x}\right] \quad for\, j \neq 0 \tag{8}$$

We use the five rationing categories discussed in the previous section, and take price rationed with a loan (prwl) as the base one. The non-random component of the j'th category score, $g(\beta_j, x_{ij})$, is modeled as the following linear function:

$$G(\beta_j, x_{ij}) = \beta_{j1} + \beta_{j2}T + \beta_{j3}W + \beta_{j4}W^2 + \beta_{j5}L + \beta_{j6}L^2 + \beta_{j7}TD + \beta_{j8}EDU + \beta_{j9}AGE + \beta_{j10}D1 + \beta_{j11}D2 + \beta_{j12}D3 + \beta_{j13}D4 \tag{9}$$

[55] Since there are k individual characteristics influencing the scores of the J+1 categories, there are a total of (k x J+1) parameters. A detailed solution to the identification problem can be found in Greene (2003) p. 721.

The variables included in the regression and their means are summarized in Table 4.7. As discussed before, we use the value of consumer durables as an indicator of household wealth (W) and a potential signal of repayment capacity for lenders. The total amount of land owned by the household (L) is measured in 'equivalent hectares' adjusting the size of parcels without irrigation and parcels under natural pastures. The inclusion of the square terms of these two variables will capture potential non-linear effects on the probability of being in a given rationing regime. In terms of wealth, for example, we expect it to have a positive effect on the probability of being price rationed with a loan, but this effect might decrease for farmers with the highest wealth levels as they will probably prefer to self-finance their investments. These farmers might be categorized as price-rationed without a loan instead.

Education and age of the household' head are included to control for individual characteristics which may influence supply and demand. Age can be considered an indicator of management experience, but it can also reflect a higher willingness to bear risk and make new investments of younger individuals. Education of the household head works as a proxy for human capital and tends to be positively correlated with farmer's productivity. Farmers with higher education levels may be also less limited in confronting the legal requirements for formal loans applications. The regional dummy variables are included to capture unobserved regional variation in demand and especially in supply conditions. The larger presence of cash crops, and the new investments in exportable crops in the Coast region should raise the probability of price rationing relative to non-price rationing mechanisms for farmers in this domain.

Table 4.7. Description of independent variables in the multinomial logit estimation.

Variable name	Description	Sample mean
T	Dummy variable for titled and registered households	0.50
W	Household's consumer durables at year 2000 in thousands of soles	0.94
W2	Household's consumer durables squared	23.89
L	Farm size in hectares	3.16
L2	Farm size squared	33.50
EDU	Years of education of household's head	5.25
AGE	Age of household's head	59.34
TD	Percentage of parcels tilted and registered in the district	0.26
D1	Dummy for region North Coast	0.29
D2	Dummy for region Central-South Coast	0.10
D3	Dummy for region North Andean	0.23
D4	Dummy for region Central Andean	0.23

4.4 Estimation results

The original results of the multinomial logit regression are reported in the Appendix 4.1. Table 4.8 presents point estimates for the marginal effects calculated at the sample means of the regressors. Standard errors were calculated using the Deltha method and are reported in parenthesis.

The first interesting result of this estimation is that Titling only seems to have a statistically significant marginal effect on the probability of being in the transaction cost-rationing category. A titled farmer who has the sample mean for all other regressors is 4 percent less likely to be transaction cost rationed than a farmer without a title., Besides this reduction in the likelihood of being non-price rationed, having a registered title does not generate a large difference for the average farmer in terms of getting access to a formal loan. This probability, however, increases with farmer's wealth and farm size. As we can see in the last column, and additional thousand soles will increase the probability of getting access to formal credit by 1.3 percent for the average farmer, while giving him one extra hectare of land will increase this probability by 0.9 percent. Education has also an important and relatively high impact on credit access: three more years of education of the head of household's is equivalent to doubling the average household wealth.

The results for the probability of a farmer being quantity-rationed are basically the opposite of the ones presented above. Having more wealth, a larger farm size, and extra years of education, decreases the probability of being part of this category. In this case, however, an additional thousand soles decreases the probability of being quantity rationed by 2.5 percent, whereas the impact of increasing the farm size in one hectare is 1.5 percent.

Table 4.8. Marginal effects from multinomial logit.

Variables	rrat		tcrat		qrat		prwol		prwl	
T	0.00376	(0.021)	-0.0448**	(0.021)	0.0317	(0.036)	-0.00743	(0.036)	0.0168	(0.013)
W	0.00963	(0.013)	0.00804	(0.009)	-0.0245*	(0.015)	-0.00588	(0.013)	0.0127***	(0.005)
L	0.00217	(0.008)	-0.00506	(0.005)	-0.0152*	(0.009)	0.00946***	(0.004)	0.00859**	(0.004)
EDU	-0.00337	(0.003)	-0.00367	(0.003)	-0.0112**	(0.005)	0.0143***	(0.005)	0.00389**	(0.002)
AGE	-0.00046	(0.001)	-0.00062	(0.001)	-0.00084	(0.001)	0.00228*	(0.001)	-0.00036	(0.000)
TD	-0.0572	(0.070)	-0.0381	(0.062)	-0.105	(0.120)	0.236**	(0.120)	-0.0361	(0.040)
D1	-0.0368	(0.032)	-0.100***	(0.024)	-0.0445	(0.064)	0.0546	(0.065)	0.127**	(0.064)
D2	-0.0812***	(0.029)	-0.0434*	(0.025)	-0.0368	(0.085)	-0.0106	(0.081)	0.172	(0.110)
D3	0.046	(0.040)	-0.012	(0.024)	-0.0259	(0.061)	-0.0127	(0.062)	0.00464	(0.039)
D4	-0.00257	(0.034)	-0.0593***	(0.021)	0.0695	(0.064)	-0.0666	(0.059)	0.059	(0.051)

Standard errors between parentheses
*** $p<0.01$, ** $p<0.05$, * $p<0.1$

Despite the relationships observed in Table 4.3 between the level of Titling Density and the percentage of farmers in each rationing category, the results of the multinomial logit regression show that after controlling for other farmer's characteristics, the impact of this variable is only statistically significant for explaining the probability of being price-rationed without a loan. This result may imply that even though formal credit supply is increased in areas with higher levels of land formalization, the loan conditions are still not attractive enough for the mean farmer.

Finally, almost none of the variables included in our model have statistically significant effects on the probability of being risk rationed. This could be due to problems in the methodology for classifying farmers into rationing regimes, or an incorrect specification of the model.

To better understand the effects of Titling on these credit rationing mechanisms we need to further explore the potential difference between titled and non-titled farmers for different levels of wealth and farm size. Therefore, we calculate the predicted probability that a titled farmer with characteristics is in the rationing regime *j*, as a function of wealth and farm size, and compare the results with the same calculation for non-titled farmers. Figure 4.4 presents the predicted probabilities of being price rationed with loan (prwl) and quantity rationed (qrat) for titled and non-titled farmers under different levels of wealth. The rest of the variables were held at their sample means.

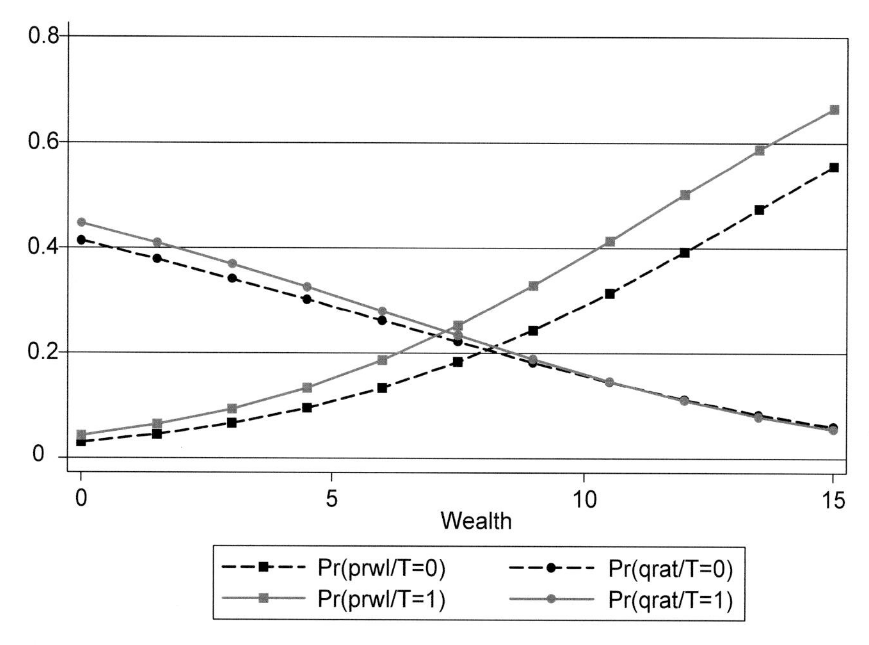

Figure 4.4. Predicted probabilities by wealth.

As expected, the probability of being quantity-rationed in the formal credit market is decreasing in household's wealth, while the likelihood of getting a loan from this sector is higher for richer farmers as compared to poorer ones with similar characteristics. However, it can be noticed that this latter effect is not exactly linear in wealth, and becomes more pronounced after it passes certain level. Considering that the mean value of this variable for the entire sample is close to 1,000 soles (or 1 in the Figure scale), our calculations imply a strong limitation for poor and even middle-income farmers in accessing these types of loans. What is more, the difference between titled and non-titled farmers reveals that the effect of titling is almost non-existent for households in the lowest part of the wealth distribution, and it only seems to make a difference for richer farmers.[56] The result for the predicted probabilities by Farm Size is presented in Figure 4.5 and shows a very similar pattern.

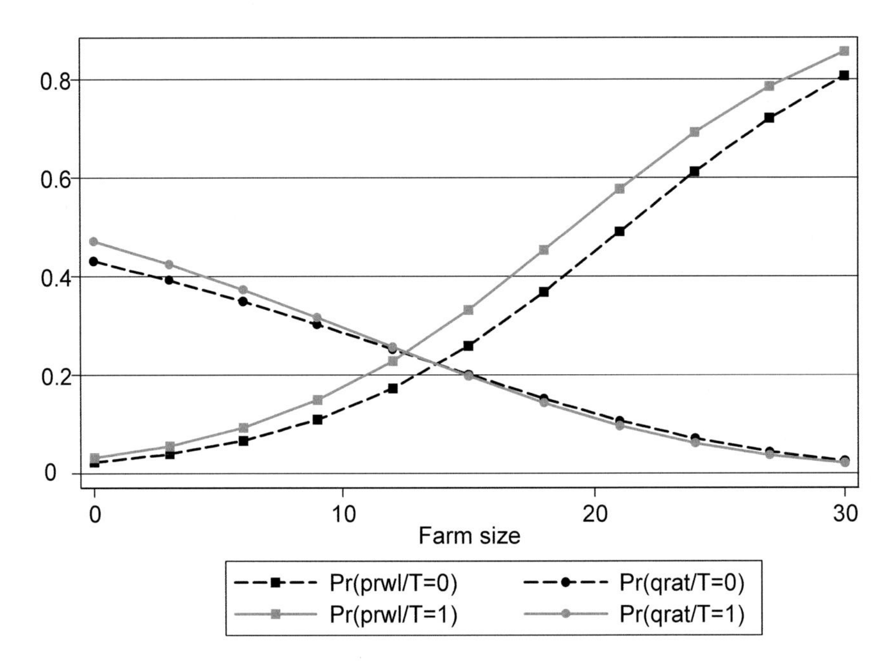

Figure 4.5. Predicted probabilities by farm size.

4.5 Concluding remarks

This chapter provides important information to understand the weak performance of the formal credit market in rural areas of Peru, and in particular the failure of the land titling policy to improve this situation. Even though formal credit institutions seem to be increasing their operations in areas with higher levels of titling density, because of an increase in their

[56] The 95% confidence intervals computed for the predicted probabilities of title and non-titled farmers become wider with increasing levels of wealth, what will make the estimates difference not statistically significant.

potential demand for loans and the higher 'liquidity' of land markets, there are other important limitations for farmers to access these loans even in these areas. The analysis of these restrictions from the farmer's perspective reveals some important features to consider in terms of policy recommendations.

More than half of our sample of farmers is non-price rationed in the credit market, with the highest percentage reporting a lack of sufficient amounts of land and collateral value as main reason for being rejected or self-excluded. The rest of them withdraw from participating in this market, even though they would have liked to apply for loans, either because they perceive the contract as bearing too much risk for them, or they lack proper information and fear the high transaction cost embedded in the application process.

Even though land titling is supposed to increase the collateral value for land and thus reduces the probability of a farmer being quantity rationed, we do not find support for this effect. Having a registered title, however, appears to reduce the transaction cost involved in formal loan applications as a result of the administrative simplification process included in the program. Still, many of these transaction costs have to be paid at the moment of application (before approval), a fact that could be discouraging many farmers from applying. These costs could be reduced or postponed, while maintaining more effective requisites to discriminate between good and bad projects instead of farmers who can afford the application cost and those who can not.

Although the regression results cannot accurately predict the characteristics of farmers under the risk-rationing category, the presence of this mechanism reveals the importance of other limitations that farmers face in the credit market. The possibility for farmers to acquire some type of insurance contract that could assist them to confront the usual negative shocks in agriculture has to be explored as a complementary policy of titling. While conventional agricultural insurance schemes might maintain adverse selection and moral-hazard limitations, new experiments on area-based yield insurance have to be explored. Other options that increase the diversification capacity of small producers could also contribute to improve their risk-bearing profile.

Finally, the existence of all these limitations implies that getting access to a formal loan becomes almost exclusively an option for wealthier farmers, with large amounts of land, and high levels of education. Land titling could facilitate their access and probably improve the conditions of their borrowing contracts, but it does not affect the possibilities for small-scale and poor producers.

Appendix 4.1. Multinomial logit coefficients estimates

	rrat		tcrat		qrat		prwol	
T	-0.309	(0.339)	-0.843**	(0.355)	-0.267	(0.289)	-0.366	(0.286)
W	-0.168	(0.178)	-0.168	(0.157)	-0.320***	(0.124)	-0.276**	(0.121)
W^2	0.007	(0.012)	0.009	(0.008)	0.012	(0.008)	0.011	(0.008)
L	-0.154	(0.120)	-0.233**	(0.107)	-0.212**	(0.090)	-0.148*	(0.089)
L^2	0.003	(0.008)	0.008	(0.006)	0.009	(0.005)	0.008	(0.005)
EDU	-0.112**	(0.043)	-0.121***	(0.045)	-0.107***	(0.035)	-0.039	(0.034)
AGE	0.003	(0.013)	0	(0.014)	0.005	(0.011)	0.014	(0.011)
TD	0.19	(1.091)	0.303	(1.116)	0.479	(0.909)	1.409	(0.908)
D1	-2.105***	(0.733)	-3.148***	(0.757)	-1.829***	(0.663)	-1.556**	(0.663)
D2	-2.852***	(0.933)	-2.278***	(0.815)	-1.737**	(0.720)	-1.673**	(0.720)
D3	0.3	(0.839)	-0.235	(0.832)	-0.157	(0.801)	-0.129	(0.807)
D4	-0.944	(0.745)	-1.745**	(0.751)	-0.754	(0.691)	-1.122	(0.699)
Constant	2.708**	(1.108)	3.590***	(1.128)	3.676***	(0.980)	2.266**	(0.977)

Log likelihood = -1113.8163 Observations: 840

Standard errors in parentheses

* significant at 10%; ** significant at 5%; *** significant at 1%

5. The externality effect of titling on investments: evidence from Peru[57]

Abstract

Many studies regarding the effects of land property rights on rural livelihoods consider individual titling as a sufficient condition for enhancing investment opportunities and increasing the value of land in the market. The empirical literature presents, however, a vast amount of evidence that challenges the principal pathways through which titling is supposed to work, and frequently relates its failure to malfunctioning of related rural markets. This paper explores yet another possible impact of titling on individual investments and land values derived from an externality effect that appears with an increase of the number of titled plots in the same district. We believe this is an important, and usually overlooked, condition for the correct functioning of the credit and land markets. Using a sample of Peruvian farmers we find that this effect indeed exists and is important for understanding the relationship between titling, investments, and land values. This result may call for the introduction of a new regional perspective in the promotion of land titling programs and complementary policies to improve the livelihoods of the rural poor.

5.1 Introduction

Land titling programs are usually based on the supposition that full-fledged land rights provide incentives and opportunities for individual farmers to invest in improved resource use strategies. Current approaches devote little attention to the importance of scale in titling and to the potential role of externalities for the development of local factor markets. This chapter therefore explores the implications of a new possible impact of titling on land investments and land values derived from an 'externality effect' that emerges with an increase in the number of titled plots in the same district (titling density).

The potential effects of land titling on the willingness as well as on the ability of farmers to invest in their land are closely related to the functioning of other markets, most notably on the markets for land sales and credit (Feder and Feeny, 1991; Binswanger *et al.*, 1995; Deininger and Binswanger, 1999). Even though titling is supposed to facilitate land transactions by reducing the cost of exchange on the land market, and improves credit access by giving land a collateral value, several studies find that individual titling does not seem to be a sufficient condition for these markets to develop or work properly (Collier, 1983; Carter *et al.*, 1994; Lopez, 1996; Carter and Olinto, 2003; Boucher *et al.*, 2004).

[57] Chapter based on joint article (with R. Ruben and J.Escobal) presented at the European School of Institutional Economics-ESNIE, Corsica 2006; and at the 11th Annual Meeting of the Latin American and Caribbean Economic Association-LACEA, Mexico City 2006.

We believe that one important condition for improving the functioning of these markets that has been overlooked in the literature, has to do with the need to count with sufficient density of formalized land rights in the area where the parcels are allocated. By reducing overall transaction costs, titling density might help to improve the dynamics of land markets and affects investment's incentives via two different channels.

First, if land markets expand as a result of an increasing general level of tenure security in the region, land becomes a more 'liquid' asset and hence any improvement made through investments can be easier realized when land is transacted. Some authors, like Besley (1995) and Platteau (1996) discussed this potential effect, but only linked it to the superior transfer rights provided by titling at the individual level. However, if better trading opportunities only appear after surpassing a particular density of land formalization in the region, titled farmers located in areas that do not meet this condition will not be able to benefit from it.

Second, formal financial institutions could be more willing to locate themselves and provide loans in areas with a higher percentage of titled plots, since it will probably be easier for them to capitalize the land given as collateral in case of defaults on loans. If credit is required for making land investments, and one of the major limitations that farmers meet came from the supply side, we could expect an improvement in their ability to make investments when titling density in the area increases. As Platteau (2000) mentions, low access to credit by titled farmers may result from supply-side failures that have their origin in various imperfections, not only in the credit market itself but also in other rural factor markets, particularly in the land market. Failure to provide credit in spite of titling may arise, for example, if titled land is not considered a reliable collateral by credit providers because it is difficult to foreclose or because - if the land market is thin - it is not easy to dispose of the land in case of default (Okoth-Ogendo, 1976; Collier, 1983; Noronha, 1985; Bruce, 1986; Barrows and Roth, 1989). If more and better credit opportunities become available to titled farmers located in areas with higher levels of titling density, this could give them an additional incentive for investing in their land.

The principal aim of this chapter is to explore the effect of titling on investments while taking into account the general level of formalization of the land rights in the districts where the parcels are located. We use information from a household' survey that was collected as part of the evaluation of the Peruvian Land Titling and Registration Program, and compare the outcomes for parcels that have been already titled and registered (T&R) with parcels that will be subject to T&R in the near future. This sample of parcels is located in more than 60 different districts in the Coastal and Andean regions of Peru, which have different levels of overall tenure security. By combining this information, we will show that investment incentives are enhanced for parcels located in districts with higher levels of titling density, and that this effect is more pronounced for land investments that have a larger contribution to (the perception of) land prices. Moreover, individual titling and the level of titling density do not only affect the perception of land prices via the investment effect, but also contribute in an independent way by reducing private enforcement costs and expanding market exchange opportunities.

In the next section we address the definition of treatment and control groups as well as other data issues related to the sample of parcels used for this study. Hereafter, we discuss the analytical model and the form of the equations to estimate the impact of titling on investments and land values. Section 4 presents and analyzes the empirical results, and Section 5 summarizes the conclusion from this study and discusses implications for policy and future research.

5.2 Characteristics of the sample[58]

For the purpose of this study, we make use of the information on the year that a parcel was subject to T&R in order to divide them into 'treatment' and 'controls' groups. Treated parcels are the ones under T&R by the program during its first stage (1994-2000), while the control group is conformed by parcels Not-T&R at the time of the survey. It is important to note that - as the program intervention works simultaneously for all parcels in the same valley - in most of the cases households will have either all or none of their parcels under T&R at this period of time.

Table 5.1 shows the distribution of sample parcels over these two groups according to the year when the program started to work on the district where the parcels are located.[59] We have

Table 5.1. Distribution of sampled parcels by year of program entrance in the district.

	Not T&R	T&R	Total
1997	333	469	802
1998	177	211	388
1999	202	249	451
2000	206	159	365
2001	158	0	158
2002	55	0	55
2003	35	0	35
2004	134	0	134
Total	1,300	1,088	2,388

[58] A detailed explanation of Land Polices and Tenure Reform in Peru during the last decades, as well as the principal characteristics of the PETT Program, can be found in Chapter 3. Here we will only highlight some characteristics of the program and data collection that are relevant to understand our analysis.

[59] Information on the year of program entry in each district is not available from the program official records (only at a higher Department level). We use the respondents' information about the year in which they received the title for this matter.

1,088 parcels in the T&R group (423 households) and 1,300 parcels for the Not-T&R group (589 households). Seventy percent of the parcels in the latter groups are located in the same districts as T&R parcels (52 Districts), and thirty percent of them belong to districts targeted by the program after the year 2000 (13 Districts).

One potential problem with the inclusion of this latter group of parcels is the possibility of having some sort of program timing-bias, relating the decision to start the program in a particular District to some characteristics that could also be affecting its expected outcome. We gathered information on the timing of the program from interviews and internal documents provided by public officials, and all suggest that intervention has been exogenous to any economic characteristics of the districts. Appendix 5.1 presents information on living conditions and accessibility before the beginning of the program from the 1993 National Population Census. Districts are divided into Early Titled Districts (first title delivered before 2000) and Late Titled Districts (first title delivered after 2000), and shows that there are no significant differences between these groups for the variables included.

For the construction of the titling density variable we use a complementary database, since we need an approximation of the total number of parcels under this category and not one just based on the sample of parcels from the PETT survey. This type of information was available from the Program's offices but only at the Department[60] level, and this would probably be too broad for accurately capturing the effects that we are testing for. The best source at hand was the National Agricultural Census of 1994, from which we recovered information on tenure status for all the parcels located in the 65 Districts that are part of our sample, and we constructed the titling density variable as the percentage of parcels subject to T&R in each district. This variable represents the initial level of formalization for each district. As the work of the PETT Program builds on that initial level, we belief it represents a good indicator for testing our hypothesis. Figure 5.1 plots all Districts by their level of titling density and the year of program' entrance to show that there is no sign of correlation between them that could affect our analysis.

The descriptive statistics of the sample are reported in Table 5.2.

Forty-six percent of the parcels in the sample have received a title from the program before the year 2000. Amongst the control parcels (54 percent of total sample), 62 percent count with some type of non-registered informal document, while 38 percent do not count with any document at all or hold only informal papers.[61] The summary of the Titling Density variable indicates sufficient variability for the districts in our sample (from 0 to 65 percent of

[60] Geopolitical division in Peru has districts as the lower level. A number of continuous Districts are aggregated in Provinces, and a group of Provinces conform Departments.

[61] Non-registered formal documents include old titles issued by the Agricultural Ministry, buy-sell contracts, or some type of public deed certified by a local judge or notary. In Fort (2007) we discuss in more detail the different types of documents present before the titling program and the level of tenure security that they bring to farmers.

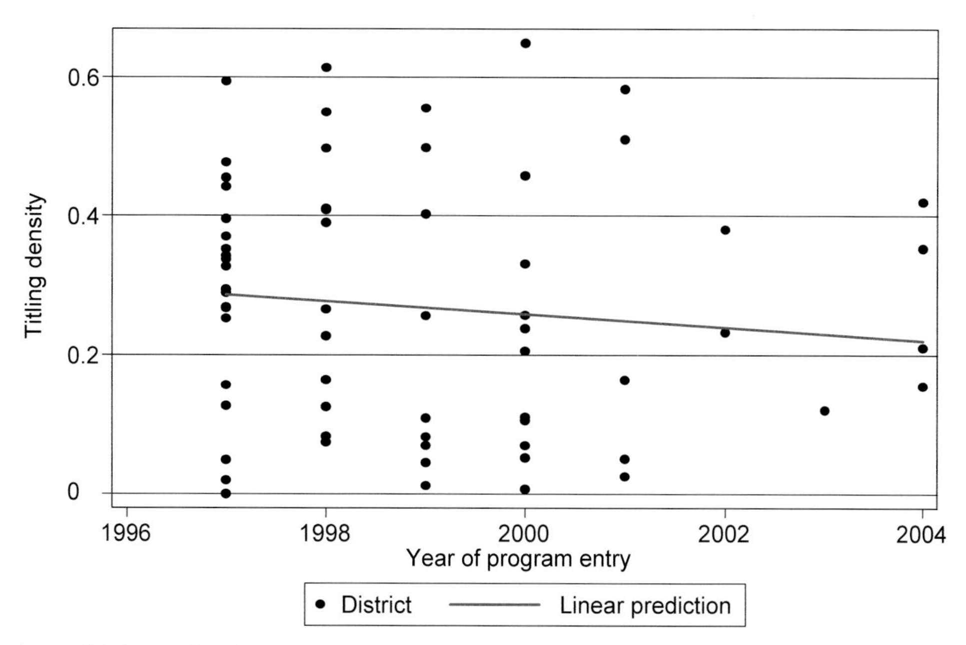

Figure 5.1. District's titling density by year of program' entrance.

parcels initially subject to T&R) and also similar mean levels for parcels in both comparison groups. Land-attached investments can be divided into infrastructure like warehouses, cattle yards, mills, drainage works, water canals, or fences; and land improvements, like terraces or land-grading. We make use of survey information on the history of investments in each parcel to divide them into investments made before and after the year 2000. Only 11 percent of the parcels in the sample report making any sort of land-attached investment in the last five years, which can be considered as a low rate since most of the farmers mention their willingness to undertake some investment in their land.[62]

Table 5.2 also includes several variables at the parcel and household level that could influence investment behavior, and as such will be included in our analysis. The mean values of these variables for the treated and control parcels in the sample reveal just a few small differences, indicating that both groups can be considered 'comparable' at these levels.

One additional piece of information that can provide some initial support for the existence of an externality effect of titling density is based on farmer's own perception of the set of rights over their land. This set included the right to choose which crop to cultivate (Use), the right to make improvements and investments in the land (Invest), the right to exclude others from using the land (Exclude), the right to inherit their land (Inherit), the right to rent-out land to

[62] For each parcel of the household, farmers were asked about their willingness to undertake any of these investments and their limitations for doing so. A positive demand was indicated for more than 80 percent of the parcels.

Table 5.2. Sample characteristics.

Variable	Total sample			T&R	Not T&R	TTEST
	Mean	Min	Max	Mean	Mean	t-value
Property rights						
Plot has a registered PETT title	0.46	0	1	1.00	0.00	-
Plot has a formal document not registered	0.34	0	1	0.00	0.62	-
Titling density	0.23	0	0.65	0.23	0.24	2.03
Investments						
Made land-attached investments after 2000	0.11	0	1	0.13	0.09	2.64
Infrastructure after 2000	0.04	0	1	0.05	0.04	1.94
Improvements after 2000	0.08	0	1	0.09	0.06	2.26
Number of land-attached investments before 2000	0.27	0	4	0.29	0.25	1.59
Number of infrastructure before 2000	0.17	0	3	0.19	0.15	2.51
Number of improvements before 2000	0.10	0	2	0.10	0.10	0.66
Parcel characteristics						
Plot size (has)	1.13	0	46.6	1.21	1.07	1.50
Erosion index (0 no problem, 3 strong erosion)	0.34	0	3	0.37	0.32	2.09
Slope index (0 no problem, 2 pronounced)	0.65	0	2	0.67	0.64	1.05
Soil quality index (1 very bad, 5 very good)	3.17	1	5	3.18	3.16	0.92
Time from parcel to district's capital (hours)	1.10	0	10	1.00	1.19	3.52
House located in the parcel	0.10	0	1	0.09	0.10	0.90
Road access to parcel (1 if paved road)	0.39	0	1	0.39	0.39	0.02
Land value (soles)	10,117	10	330,000	9,245	10,908	1.74
Household characteristics						
Sex head of household	0.84	0	1	0.86	0.82	2.73
Age head of household	58.84	24	98	58.59	59.05	0.84
Mean years of education adults	6.59	0	16	6.54	6.64	0.59
Spanish main lenguage	0.65	0	1	0.60	0.68	4.04

others (Rent Out), and the right to sell land (Sale). Farmers were first asked to mention how secure they feel about each of these rights (Secure/Insecure/Indifferent) and after that to state the strength of that level of security (Fully/Very much/Not much). Using these answers we were able to construct a scale from 1 to 7 for the perception of each right (1=totally insecure, 7=totally secure). Appendix 5.2 presents the mean values of the rights perception index (1-7) for T&R and Not-T&R households, and shows that in all cases T&R farmers show higher security perception, with statistical significant differences for the rights to Use, Invest, and Inherit. Even more important for our analysis is the relationship between these security' perceptions and the levels of titling density in the Districts where farmers are located.

As we can see in Figure 5.2, the whole set of right's perceptions increases for T&R farmers as the level of titling density increases, while there does not seem to be any particular correlation for the Not-T&R farmers.

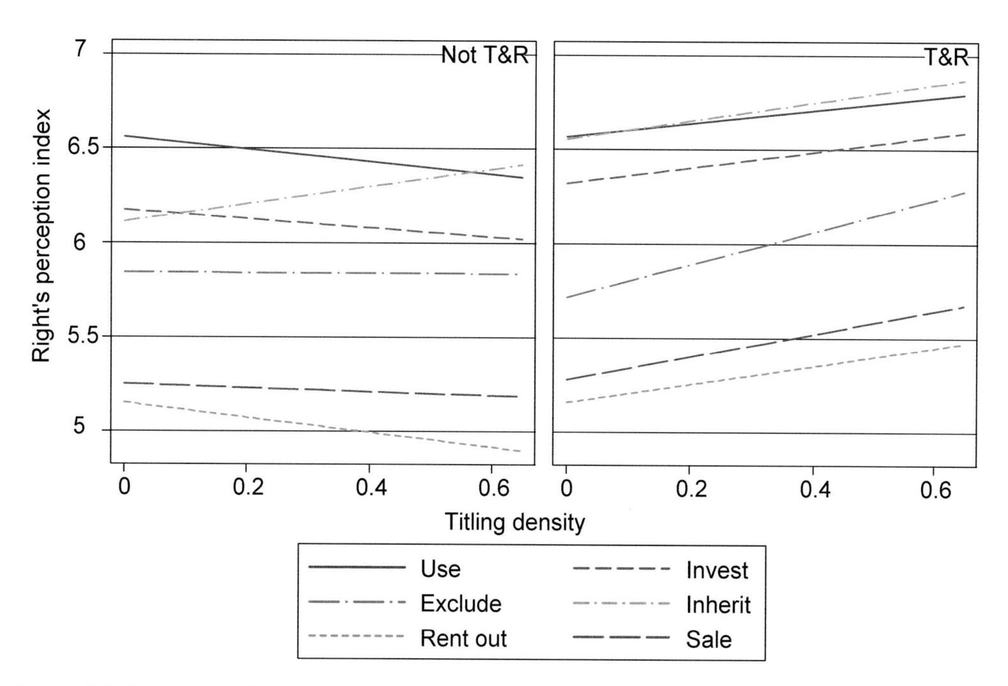

Figure 5.2. Perception of rights and titling density (linear predictions by group).

5.3 Analytical framework

In this section we introduce the framework for the analysis of the 'externality effect' of titling on investments, and for the relationship between tenure security and land values. Our main interest is to identify whether the overall level of formalization of land rights in a district, measured by the titling density variable, produces an effect on the probability of making land related investments. For that matter, we include this variable in the traditional equation framework to explain investments at the parcel level in the following way:

$$I^k_{ij} = \alpha T_{ij} + \beta X_{ij} + \delta Z_i + \gamma D_i + \tau(T_{ij} * D_i) + e_{ij} \qquad (1)$$

Where I^k_{ij} represents investments of type k undertaken by farmer i in parcel j during the period 2000-2004, in such a way that we only consider decisions made after the titling period selected for this study. The effects of property rights are measured by T_{ij} which separates parcels into T&R, Not-T&R but with a formal ownership document, and Not-T&R without this type of document. The vector X_{ij} includes several parcel-specific characteristics that could be affecting the suitability of the land for different types of investments, while Z_i includes some characteristics of the household. One variable of particular importance in X_{ij} is the number of past investments made in each parcel (up to the year 2000), what makes the estimation if I^k_{ij} conditional on those previous values. Apart from picking up unmeasured parcel characteristics, this variable also depends on past values of the explanatory variables and in particular on previous levels of tenure security. Therefore, we will instrument it by using information on the different documents that farmers hold for their parcels before titling.[63]

The coefficient for D_i captures the 'externality effect' on new investments of being in a district with a particular level of titling density, and the interaction of this variable with the dummy for property rights at the parcel level (T_i* D_i) indicates if there is an additional effect of titling on investments when parcels are located in districts with higher levels of land rights formalization.

In addition to this analysis, we also want to know if land-attached investments contribute to an increase in land values and to which extent - over and above any potential impact on investments - the possession of a title and the degree of titling density also improves this measure of wealth. We use the principle of hedonic land price estimation (Rosen, 1974) by which all factors that can possibly affect the expected flow of returns to land can be capitalized in land values. Since we use self-reported information on the prices at which farmers will be willing to sell their land instead of actual sales market prices, we believe this methodology is the most appropriate procedure. A potential problem with this estimation technique may arise if there is some unobserved parcel or household characteristic that affect both land values and other explanatory variables included in the regression, thus occasioning potential biases in the coefficient estimates. The study by Deininger and Chamorro (2004) in Nicaragua argues, for example, that there might be unobserved household-specific characteristics related to wealth affecting both land values and the probability of having a title, so that failure to control for them will generate a biased coefficient for the effect of titling.[64] In our case, however, the set-up of the titling program and the construction of our treatment and control groups make this possibility very unlikely to occur. Estimation of land values then takes the form:

$$LV_{ij} = \alpha T_{ij} + (\beta_1 X_{ij} + \beta_2 I^k_{ij}) + \delta Z_i + \gamma D_i + e_{ij} \qquad (2)$$

We separate the vector of parcel characteristics to include the value of all land-attached investments made in the parcel, and measure the marginal effect that a particular type of investments has for the perceived value of land. The coefficient of individual property rights (α) and the one for the titling density variable (γ) will capture then the independent

[63] In Chapter 3 we found that the level of investments before titling was in fact related to property rights held at that time, and that the impact of titling on new investments was dependent on that previous level.

[64] In particular they claim that if poor households have higher discount rates (which affect their value of land) and are less likely to be titled, the coefficients from the hedonic regression may be biased.

effects of having more secure property rights and of being located in areas with expanded market opportunities, respectively. We also include some characteristics of the households (Z_i) that could be affecting the perception of land values, like age and gender of the head of household.

5.4 Empirical results

Equation 1 is estimated for the probability of having made an investment after the year 2000 (Probit model) and also for the value of those investments (Tobit model). Both estimations yield similar results and so we will discuss our findings based on the outcomes of the Probit model.[65] In this case, I_{ij}^{k} is a variable that equals one if household i undertook investment k in parcel i, and zero otherwise. We assume that the household undertakes the investment if the expected return from doing so is positive, so that

$$I_{ij}^{k} = \begin{cases} 1 & \text{if } \pi_{ij}^{k} > 0 \\ 0 & \text{otherwise} \end{cases}$$

This estimation is done first considering all land-related investments undertaken in parcel j, and then separating them into investments in infrastructures and investments in land improvements. We take this approach because we believe that - even though both types of investments could raise land productivity, and require land tenure security - farmer's decision to undertake them might be based on different motives that are of particular interest to our analysis. Investing in improvements like terraces or land-grading are more likely to occur on parcels facing erosion or slope problems, and are usually more labor than capital-intensive. Investing in infrastructures requires larger amounts of capital, and might produce a higher impact on the value of land in the market. As such, we expect the titling density effect to be stronger for the latter type of investments, while also producing a higher impact on the expected land values.

Results for the Probit estimation of Equation 1 are provided in Table 5.3. For each type of investments, column (1) estimates the model without the inclusion of the titling density variable, column (2) incorporates it, and column (3) adds the interaction between individual titling and the titling density variable. As recent econometric studies suggest that in non-linear models the magnitude of the interaction effect is different from the marginal effect of the interaction term (Norton *et al.*, 2004), we will contrast the (τ) coefficient from the Probit estimation with the results of applying the method suggested by Norton *et al.* (2004)[66], and also with the results from a linear probability model[67].

[65] Estimation results for the Tobit model are shown in the Appendix 5.3.

[66] Because the sign and values of the interaction effect computed by this method might be different for different values of the explanatory variables, we report in the Appendix 5.5 the sample average for these parameters.

[67] This estimation is based on Wooldridge (2002) pg. 454: The Linear Probability Model for Binary Response, who suggests using a weighted least squares regression.

Table 5.3. Probability of having made land-attached investments.

	All land-attached Investments			Infrastructure			Improvements		
	(1)	(2)	(3)	(1)	(2)	(3)	(1)	(2)	(3)
Registered PETT title	0.053***	0.056***	0.054**	0.031***	0.030***	0.009	0.026*	0.028**	0.037*
	(3.13)	(3.26)	(2.23)	(3.00)	(2.91)	(0.63)	(1.96)	(2.10)	(1.91)
Formal Document not registered	0.032*	0.034*	0.034*	0.019	0.016	0.017	0.012	0.012	0.012
	(1.74)	(1.80)	(1.80)	(1.64)	(1.43)	(1.50)	(0.82)	(0.83)	(0.83)
Number of investments before 2000	-0.207*	-0.204*	-0.205*	-0.032	-0.026	-0.038	-0.326**	-0.282**	-0.281**
	(1.79)	(1.79)	(1.79)	(0.40)	(0.34)	(0.50)	(2.52)	(2.19)	(2.18)
Plot size	0.008***	0.007***	0.007***	0.000	0.000	0.000	0.005***	0.005***	0.005***
	(3.08)	(3.03)	(3.03)	(0.11)	(0.07)	(0.17)	(3.27)	(3.14)	(3.12)
Erosion index	0.030***	0.030***	0.030***	0.010**	0.010**	0.010**	0.023***	0.022***	0.022***
	(3.77)	(3.79)	(3.79)	(2.35)	(2.37)	(2.36)	(3.45)	(3.39)	(3.39)
Slope index	0.020**	0.022**	0.022**	-0.000	-0.001	-0.001	0.024***	0.026***	0.026***
	(2.01)	(2.21)	(2.21)	(0.00)	(0.24)	(0.12)	(2.88)	(3.06)	(3.05)
Soil quality index	0.017	0.022*	0.022*	0.004	0.004	0.005	0.013	0.017*	0.017*
	(1.46)	(1.88)	(1.88)	(0.67)	(0.66)	(0.75)	(1.40)	(1.75)	(1.75)
Time from parcel to district's capital	-0.004	-0.003	-0.003	0.002	0.002	0.002	-0.004	-0.004	-0.004
	(0.82)	(0.72)	(0.72)	(1.11)	(1.18)	(1.07)	(0.99)	(0.86)	(0.85)
Road access to parcel	-0.041***	-0.033**	-0.033**	-0.005	-0.006	-0.006	-0.028**	-0.018*	-0.019*
	(2.90)	(2.33)	(2.33)	(0.69)	(0.74)	(0.78)	(2.57)	(1.72)	(1.77)
Sex head of household	0.038**	0.032*	0.032*	0.020**	0.020**	0.019**	0.018	0.010	0.010
	(2.34)	(1.92)	(1.92)	(2.08)	(2.14)	(2.09)	(1.40)	(0.72)	(0.73)
Age head of household	-0.002***	-0.002***	-0.002***	-0.001***	-0.001***	-0.001***	-0.001***	-0.001***	-0.001***
	(4.21)	(4.09)	(4.10)	(3.68)	(3.84)	(3.81)	(3.82)	(3.65)	(3.64)
Mean years of education adults	0.005***	0.005***	0.005***	-0.000	-0.000	-0.000	0.006***	0.006***	0.006***
	(2.85)	(2.76)	(2.76)	(0.40)	(0.16)	(0.12)	(4.32)	(4.14)	(4.14)
Spanish main lenguage	0.053***	0.055***	0.055***	0.018	0.021**	0.023**	0.044***	0.040**	0.039**
	(2.74)	(2.80)	(2.79)	(1.61)	(2.01)	(2.19)	(2.74)	(2.50)	(2.44)
Titling density		0.091*	0.087		0.055**	0.015		0.044	0.061
		(1.92)	(1.50)		(2.05)	(0.45)		(1.16)	(1.31)
Titling density* T&R			0.008			0.079**			-0.033
			(0.13)			(2.14)			(0.61)
Observations	2,388	2,388	2,388	2,388	2,388	2,388	2,388	2,388	2,388
Pseudo R-squared	0.09	0.10	0.10	0.09	0.10	0.10	0.10	0.12	0.12

Robust z statistics in parentheses
* significant at 10%; ** significant at 5%; *** significant at 1%
Coefficients correspond to marginal probabilities at the mean values.
Regional dummies and control variables at District's level included but not reported.

The Probit model results for the decision of undertaking any type of land-attached investments suggest that having stronger land rights increases the investment probability. Having a registered title from the PETT program raises the probability of investing by more than 5 percent in comparison with a parcel without any documentation or with only informal documents. Parcels with formal but not registered documents have 3 percent more chance of presenting investments, but this coefficient is only significant at the 10 percent level. These results are confirmed by the Tobit regression (see Appendix 5.3) for the value of investments, where the conditional marginal effects of having a registered title is almost two times the effect of having a formal non-registered document.

Having made investments before seems to reduce the probability of pursuing new investments. As we mentioned before, this variable depended on previous levels of land rights and so it was instrumented for using this information. In Chapter 3 we already noticed that before the titling program farmers already held different types of documents to prove possession or ownership over their parcels, and that there was a relation between stronger documents and investment levels. Moreover, we demonstrated that the effect of titling on new investments was stronger for parcels with previous lower levels of tenure security. As such, the negative coefficient found here for this variable might be picking up that relationship.

Investments are more likely to be made on larger parcels, and also on parcels facing some erosion and slope problems. In terms of household characteristics, male, younger, and Spanish-speaking heads of households are more likely to make new investments than their counterparts, and also higher levels of education among adults in the family seem to induce this behavior. The inclusion of the titling density variable in the regression (column 2) does not modify substantially previous results, and shows a positive effect (even though only significant at a 10 percent level) of a 9 percent increase in the probability of making investments for a unitary change on titling density. Considering that titling density was constructed as a percentage between (0,1), a better interpretation of this effect can be given in terms of the increase in one standard deviation from its mean (0.17). Thus, having a parcel in a district where titling density exceeds the mean by that amount increases the probability of investing on it by 1.5 percent. The addition of the interaction between individual titling and the level of titling density (column 3) generates a positive but not-significant coefficient. This result is confirmed by the linear probability model (Appendix 5.4) and by the average estimate of the interaction effect (Appendix 5.5) using the method by Norton *et al.* (2004).

Estimation of Equation 1 for the two different sets of investments yields very interesting results for our analysis. First, individual titling and registration have a positive and similar effect for both types of investments in terms of the increase on the probability to make investments, but it differs when considering the value of investments. The Tobit regression reports an increase of S/. 69.5 on the value of infrastructure investments while the increase for land improvements is only of S/. 24. Second, investing in land improvements is much more likely to happen in parcels with erosion and slope problems, and also in more isolated parcels. Finally, the addition of the titling density variable (column 2) and its interaction with individual titling (column 3) have a positive and significant effect for the regression results on infrastructure, but no significant effect was found for the estimation of land improvements. The results for the interaction term are maintained under the linear probability model (Appendix 5.4), but the calculations using the Norton *et al.* (2004) method cast some doubts on the robustness of the interaction effect for infrastructure (Appendix 5.5).[68]

Increasing titling density by one standard deviation raises the probability of investing in infrastructure by approximately 1 percent for all parcels in the district. The coefficient of the Probit model on the interaction term implies that the marginal effect of individual titling on the probability of having made an investment of this type increases by 1.3 percent for every additional standard deviation from the mean of titling density. Based on the results from the Probit model (column 3), Figure 5.3 shows how the predicted probability of investing in infrastructure increases with the level of titling density, but only for T&R parcels.

Table 5.4 reports the results for the estimation of equation 2. As we can see, land related investments do in fact contribute to an increase in land values and, as we hypothesized before,

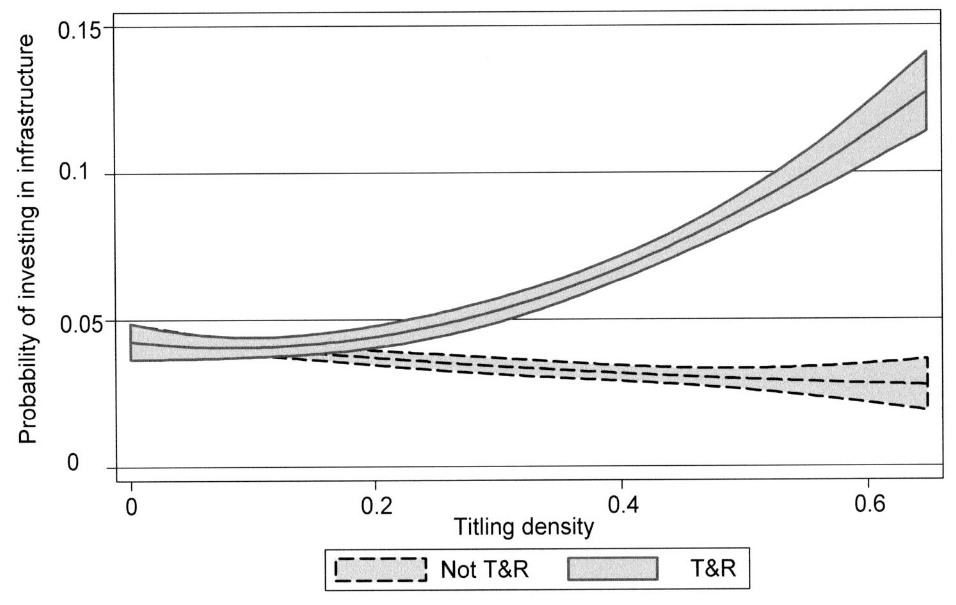

Figure 5.3. Predicted probability of investing in infrastructure, by group and titling density.

[68] Appendix 5.6 compares the marginal effect from the Probit model with the interaction effect computed with the Norton *et al.* (2004) method for different values of the predicted probability of investing in infrastructure. Even though they seem to be strongly correlated, the standard errors reported with the new computation are extremely high and consequently z-values are very low.

Table 5.4. OLS regression on land values (log).

Variables	Coefficients	t-stat
Registered PETT Title	0.336***	(4.25)
Formal document not registered	0.290***	(3.56)
Titling density	1.063***	(5.91)
Population pressure at 1993	0.316***	(5.99)
Value of infrastructure (log)	0.081***	(6.72)
Value of improvements (log)	0.046***	(3.48)
Plot size	0.204***	(8.76)
Time from parcel to district's capital	-0.049**	(2.09)
Erosion index	-0.153***	(3.64)
Slope index	0.079*	(1.86)
Soil quality index	0.328***	(7.07)
Road access to parcel	0.509***	(8.27)
House located in parcel	0.158*	(1.74)
Sex head of household	0.005	(0.07)
Age head of household	0.009***	(4.28)
Mean years of education adults	0.039***	(5.55)
Spanish main lenguage	0.334***	(3.88)
Constant	4.654***	(18.32)
Observations	2.268	
R-squared	0.43	

Robust t statistics in parentheses
* significant at 10%; ** significant at 5%; *** significant at 1%

the marginal effect of investing in infrastructures almost doubles the effect of investing in improvements.[69] Apart from its indirect impact through enhanced investments, having a registered title increases the value of land in almost 34 percent as compared to the value of parcels without any formal documentation. The coefficient for having a formal but not registered title reveals that these types of documents also provide tenure security to farmers and could facilitate land transactions by reducing enforcement costs. However, as we have shown in Chapter 3, the acquisition of these documents might be limited for poorer farmers, and therefore public titling policies remain as a key option for these types of households.

[69] The F test for this difference result in a value of 3.64, so that the null hypothesis of the coefficients being equal was rejected with a 95% probability.

At the District level, we also included a measure of population pressure from the 1993 National Census (rural inhabitants/agricultural area) to verify if this could be affecting demand for land and hence land prices. The coefficient for this variable confirms that the value of land is higher for districts that are more densely populated. Finally, our results indicate that, on average, land values are higher in districts with higher levels of titling density. According to the coefficient for this variable, a 1 percent increase in the level of titling density will increase average land values in that Districts by a similar percentage, which represents a very important impact of titling policies in rural areas.

5.5 Concluding remarks

The evidence presented in this chapter suggests that - although individual titling and registration can increase the level of farmer's tenure security and as such contribute to the enhancement of land related investments in their parcels - the strength of this relationship is very much related to the density of land rights formalization in the area where parcels are located. This factor is of particular importance for stimulating investments that have a high contribution in raising the value of land.

We also showed that, apart from its effect on investments, individual title and registration and the level of titling density have a positive impact on farmer's perception about the price of their land. Individual T&R reinforces tenure security and helps reducing transaction and enforcement costs, while increased levels of titling density reflect expanded market opportunities for land.

These findings indicate, on the one hand, that for individual titling policies to become effective, other conditions to reduce transaction costs in rural areas and to improve the dynamics of land markets need to be fulfilled, while on the other hand, these conditions can at least be partially improved when the levels of titling density start to increase. However, this policy alone might not be sufficient for lifting up other important constraints in rural markets. The relative small change in the probability of investment, and on the investment value, reveals that there is still a positive demand for land-related investments (in particular for infrastructures) amongst many farmers that cannot be satisfied mostly because of a lack of financial opportunities. When asked about their limitations, around 70 percent of the farmers revealing a positive demand mentioned the lack of credit as their principal constraint to undertake investments. Therefore, even if formal financial institutions would be more willing to locate themselves and provide credits in Districts with higher levels of titling density, binding constraints from the demand side, as the lack of information or insurance against negative shocks, and the extremely small size of land-holdings, need to be addressed in order to enhance the potential benefits of this intervention. A more detailed study regarding the operation of the financial market is therefore of critical importance (see Chapter 4).

In terms of the effect of individual titling and titling density on reducing transaction and enforcement costs in the market for land sales, more work needs to be done in order to investigate the pattern of actual transactions and its implications. If financial opportunities are biased towards wealthier farmers, it could be easier for them to accumulate land in these newly developed markets. This will not necessarily be an inefficient outcome, considering the extreme land fragmentation in many regions of Peru, but it could yield negative effects in terms of inequality if poorer farmers recur to distress sales as a way of cooping with negative shocks, and also if local off-farm opportunities are limited. Policies that facilitate access to land for small but efficient farmers (or groups of farmers), and that assist in the development of insurance mechanisms for them, could be essential complements for the Land Titling Program.

Appendix 5.1. District characteristics at 1993

	Population	Sewerage*	Electricity*	Paved Road^
Early	7,443	38%	24%	25%
Late	7,378	34%	23%	26%

* Report the mean of the % of households with access in the District
^ Reports the % of Districts where principal access is a Paved Road

Appendix 5.2. Perception of Rights by groups

	Not T&R	T&R	t-value
Use	6.48	6.66	2.81
Invest	6.11	6.43	3.59
Exclude	5.84	5.94	0.85
Inherit	6.24	6.68	5.43
Rent Out	5.05	5.28	1.70
Sale	5.22	5.44	1.56

Appendix 5.3. Tobit regression for the value of investments made since 2000

	All land-attached Investments			Infrastructure			Improvements		
	(1)	(2)	(3)	(1)	(2)	(3)	(1)	(2)	(3)
Registered PETT title	55.7***	58.6***	54.4**	69.5**	67.1**	27.7	24.0**	25.3**	25.1
	(2.84)	(3.00)	(2.00)	(2.53)	(2.51)	(0.76)	(2.05)	(2.14)	(1.51)
Formal document not registered	36.4*	32.7	32.7	53.7*	45.9	46.8*	4.9	2.9	2.9
	(1.68)	(1.52)	(1.53)	(1.84)	(1.62)	(1.65)	(0.38)	(0.22)	(0.22)
Number of investments before 2000	-0.4*	-0.3*	-0.3*	-0.3	-0.3	-0.3	-0.3*	-0.2	-0.2
	(1.84)	(1.71)	(1.71)	(0.68)	(0.70)	(0.75)	(1.90)	(1.49)	(1.49)
Plot size	7.9***	7.3***	7.3***	0.9	0.8	0.9	4.3***	4.1***	4.1***
	(2.76)	(2.61)	(2.61)	(0.18)	(0.17)	(0.18)	(3.13)	(3.06)	(3.06)
Erosion index	35.5***	36.2***	36.2***	24.4**	25.6**	25.5**	21.0***	20.2***	20.2***
	(3.47)	(3.57)	(3.58)	(1.99)	(2.13)	(2.12)	(3.28)	(3.16)	(3.16)
Slope index	17.2	17.7	17.8	-1.4	-4.4	-3.2	15.8**	16.2**	16.2**
	(1.52)	(1.58)	(1.59)	(0.10)	(0.32)	(0.24)	(2.13)	(2.19)	(2.18)
Soil quality index	19.0	22.7*	22.8*	18.0	17.9	18.7	5.1	7.5	7.5
	(1.48)	(1.77)	(1.78)	(1.04)	(1.06)	(1.10)	(0.68)	(1.00)	(1.00)
Time from parcel to district's capital	-4.2	-2.9	-2.9	5.3	6.0	5.6	-3.5	-2.9	-2.9
	(0.74)	(0.52)	(0.53)	(0.70)	(0.82)	(0.75)	(1.02)	(0.84)	(0.84)
Road access to parcel	-29.5*	-19.0	-18.8	-10.1	-9.8	-9.6	-12.8	-5.1	-5.1
	(1.78)	(1.14)	(1.13)	(0.44)	(0.44)	(0.43)	(1.35)	(0.52)	(0.52)
Sex head of household	46.1**	37.2*	37.1*	63.7**	63.8**	62.5**	11.2	3.4	3.4
	(2.18)	(1.77)	(1.76)	(2.07)	(2.11)	(2.07)	(0.90)	(0.27)	(0.27)
Age head of household	-2.1***	-2.1***	-2.1***	-2.4***	-2.4***	-2.4***	-1.2***	-1.2***	-1.2***
	(3.96)	(3.96)	(3.96)	(3.29)	(3.38)	(3.36)	(3.67)	(3.60)	(3.60)

	All land-attached Investments			Infrastructure			Improvements		
	(1)	(2)	(3)	(1)	(2)	(3)	(1)	(2)	(3)
Mean years of education adults	5.6***	5.5***	5.5***	-0.6	0.1	0.2	5.2***	5.0***	5.0***
	(2.70)	(2.69)	(2.69)	(0.21)	(0.03)	(0.06)	(4.03)	(3.88)	(3.88)
Spanish main lenguage	54.6**	56.8**	57.3**	38.7	50.0*	54.6*	36.5**	33.1**	33.2**
	(2.27)	(2.37)	(2.38)	(1.29)	(1.68)	(1.80)	(2.39)	(2.16)	(2.15)
Titling density		125.5**	117.0*		159.5**	85.2		45.1	44.6
		(2.29)	(1.74)		(2.33)	(1.02)		(1.30)	(1.03)
Titling density* T&R			16.9			152.5			0.9
			(0.22)			(1.53)			(0.02)
Constant	-332.1***	-297.4***	-298.1***	-392.6***	-505.2***	-514.7***	-189.1***	-134.3***	-134.3***
	(4.83)	(3.83)	(3.83)	(4.24)	(4.09)	(4.14)	(4.35)	(2.82)	(2.81)
Observations	2,388	2,388	2,388	2,388	2,388	2,388	2,388	2,388	2,388
Pseudo R-squared	0.02	0.03	0.03	0.03	0.03	0.04	0.03	0.04	0.04

Absolute value of z statistics in parentheses.

* significant at 10%; ** significant at 5%; *** significant at 1%

Coefficients correspond to the conditional marginal effects for S/. invested (E(y/y>0)).

Regional dummies and control variables at district's level included but not reported.

Appendix 5.4. Linear probability model

	All land-attached investments		Infrastructure		Improvements	
Registered PETT title	0.083***	(2.78)	0.017	(0.75)	0.072***	(2.66)
Formal document not registered	0.054**	(2.27)	0.023	(1.37)	0.028	(1.30)
Number of investments before 2000	-0.312**	(1.98)	-0.029	(0.20)	-0.425*	(1.95)
Plot size	0.010***	(3.16)	0	(0.03)	0.009***	(3.68)
Erosion index	0.035***	(3.25)	0.017**	(2.23)	0.023**	(2.35)
Slope index	0.034***	(2.61)	-0.005	(0.62)	0.042***	(3.11)
Soil quality index	0.030**	(2.01)	0.013	(1.25)	0.02	(1.50)
Time from parcel to district's capital	-0.005	(0.87)	0.004	(0.94)	-0.012**	(1.98)
Road access to parcel	-0.045**	(2.30)	-0.009	(0.65)	-0.045***	(2.78)
Sex head of household	0.026	(1.08)	0.023	(1.27)	0.003	(0.13)
Age head of household	-0.003***	(4.28)	-0.002***	(5.14)	-0.002***	(3.89)
Mean years of education adults	0.005**	(2.30)	-0.001	(0.82)	0.009***	(4.28)
Spanish main lenguage	0.084***	(3.04)	0.056***	(2.76)	0.057**	(2.17)
Titling density	0.155**	(2.06)	0.083	(1.53)	0.101	(1.45)
Titling density* T&R	-0.034	(0.39)	0.134**	(2.12)	-0.084	(1.04)
Observations	2,205		2,080		2,107	
R-squared	0.06		0.04		0.07	

Absolute value of t statistics in parentheses
* significant at 10%; ** significant at 5%; *** significant at 1%

Appendix 5.5. Interaction effects of probit model

	Investments			
	Mean	**Std. dev.**	**Min**	**Max**
Interaction effect	0.048	0.013	0.002	0.065
Standard error	1.781	2.741	0.151	14.772
z-value	0.069	0.066	0.001	0.300

	Infrastructure			
	Mean	**Std. dev.**	**Min**	**Max**
Interaction effect	0.151	0.101	0.005	0.475
Standard error	16.192	25.926	0.230	126.215
z-value	0.022	0.022	0.001	0.419

	Improvements			
	Mean	**Std. dev.**	**Min**	**Max**
Interaction effect	-0.022	0.021	-0.118	0.000
Standard error	2.382	3.297	0.015	25.243
z-value	-0.016	0.015	-0.152	0.020

Appendix 5.6. Interaction effect for infrastructure, by predicted probability

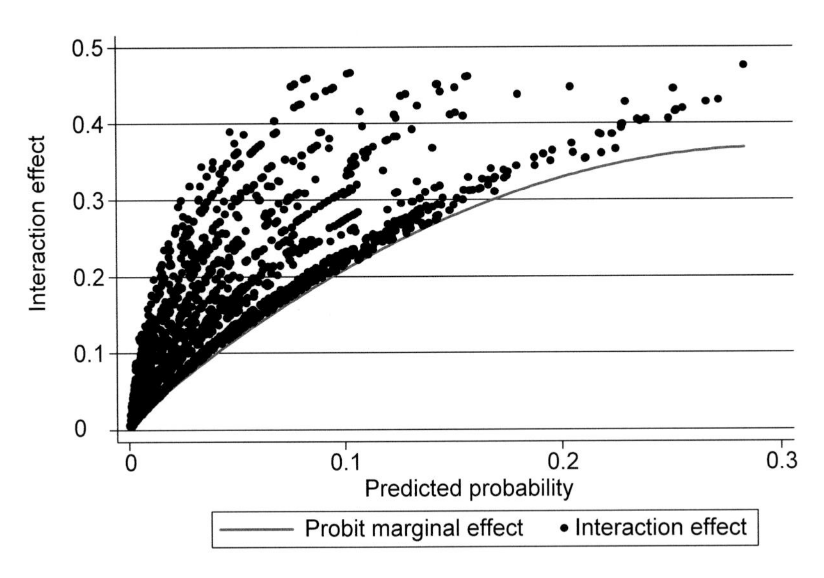

6. Discussion and conclusions

6.1 Introduction

The importance of well-defined property rights for economic progress has being implicitly recognized since the beginning of the economics profession and by many thinkers of different schools. From Adam Smith's essential proposition that economic growth largely depends on the extent of division of labor, to Marx's materialistic theory of history, property rights appear as the core institutions to discard the fear of fraud in transactions, or to legally maintain control over key production factors.[70]

However, the currently dominant neo-classical paradigm in the field of economics generated an inappropriate set of tools to deal with the subject of economic development, as it was more concerned with the functioning of markets than with the ways they evolved. It took for granted politics, demography, and institutions – the necessary building blocks for understanding the process of economic change (North, 2000). Moreover, according to neo-classical theory, if there are initial differences in individuals' skills and endowments of production factors, markets should help to optimize factor proportions employed for production activities, increasing in that competitive way the overall efficiency of resources allocation in the economy.

A general characteristic of developing countries, though, is that many of these markets are missing or work in an imperfect way. The core subject of development economics is then to analyze the economic behavior of individuals in this context and its results, not only in terms of allocative efficiency but also taking into account the potential welfare effects in terms of inequality and poverty (De Janvry and Sadoulet, 1995; Ray, 1998).

Since the emergence of the Washington Consensus, many Less Developed Countries (LDC) engaged themselves in the application of market liberalization reforms in an attempt to 'put the markets at work' in order to 'get the prices right'. Soon after these reforms began, a second generation of reforms was presented as a necessary condition for the markets to function well or even just for the development of markets where they did not exist: without 'getting institutions right' markets cannot develop or function as well as predicted (World-Bank, 1997). But putting in place the formal institutions that have under girded the growth of the developed world does not always produce the desired results (North, Platteau, 2000). That is probably so because formal rules must be complemented by informal norms of behavior (and enforcement characteristics) to get the desired results, and also the setting of these rules have to account for initial/structural socio-economic conditions as well as multiple market imperfections commonly present in LDC's.

[70] For an extended discussion of this idea see Platteau (2000).

In this context, the present study first discusses the links between land access, property rights, and economic growth; and then analyzes the results of a public intervention, Land Titling and Registration, that constitutes one of the main instruments for contemporary land policy in LDC's. With a macroeconomic perspective on the first analysis, and meso (or regional) and micro level approach for the study of the Peruvian Land Tilting and Registration Program, this study attempted to provide a more comprehensive analysis and discussion of the importance of institutions like land property rights in the context of market liberalization reforms.

This final chapter summarizes the main findings of our study and places them into a policy perspective. Section 2 presents the principal debates on the topic, which motivated our research questions in Chapter 1, and provides some answers to them in the light of our findings. Section 3 discusses the main implications and policy recommendations derived from the renewal results, and section 4 presents some limitations of the study as well as a few directions for future research.

6.2 Key debates and main findings

6.2.1 Land access, property rights, and economic growth

The relationship between land distribution, property rights, and economic growth has been largely unexplored at the macroeconomic level. Even though many studies address separately the implications of wealth inequality for economic growth, on the one side, and the importance of institutions like a well-defined property rights system on the other side, so far there is no study that jointly explores these links. Moreover, most studies exploring the relationship between inequality and growth rely on measures of income inequality rather than assets distribution as an explanatory variable. This is troublesome since the theoretical relationship between inequality and growth is better explained by assets distribution than by income.

Chapter 2 provides a theoretical discussion and empirical evidence to better understand the relationships between land distribution and economic growth when the role of secure property rights is accounted for. Using an innovative panel data set with changes in land distribution over time and between countries, we have been able to provide confirmation for the hypothesis that asset distribution is a major determinant of economic growth. Apart from a direct effect, we also show that land inequality creates a barrier to the effectiveness of educational policies, confirming initial findings of Deininger and Olinto (1999). Moreover, the incorporation of the physical investments variable in the model corroborates the existence of a growth-reducing impact of land inequality that goes beyond the conventional channel of credit market imperfections and reduced investments.

Even though the security of property rights appears as an important factor to explain economic growth, its effect does not modify the relationship found between land inequality and growth, as Keefer and Knack (2002) argued. The omission of the investment variable in their model is the likely reason for this discrepancy. Future research needs to incorporate the potential relationship between property rights and investments in order to clarify their individual influence over the interactions between land inequality and economic growth.

6.2.2 Land rights formalization, tenure security, and investment' incentives

The lack of tenure security over land is widely recognized as an important limitation for farmers to maximize the potential returns of this resource. Unclear definition of individual property rights can give rise to disputes over ownership, inheritance, or land boundaries. Ultimately, it increases the probability of losing the land in a dispute, and with this all effort and investments devoted to it.

Even though land titling programs are fundamentally promoted to increase farmer's tenure security and investment incentives, the justification for this type of public intervention is increasingly questioned on the grounds of its limitation to replace or improve the effect of informal or customary rights already in place. Some authors argue that in many rural areas, customary rights provided by local authorities, or farmer's acquisition of informal land documents, might be sufficient to provide them with the required tenure security to induce investments (Migot-Adholla, 1991; Platteau, 1992; Bruce and Migot-Adholla, 1994).

The results in Chapter 3 suggest that land titling policies aiming to formalize individual land rights have a differentiated effect on investments, depending on the farmer's level of tenure security over a parcel before the policy initiates. We show that before the intervention of the program, parcels can be already categorized into different levels of tenure security depending on the type of informal documents held by farmers. Accordingly, parcels with 'stronger' documents present initially higher levels of investments compared to parcels with 'weaker' documents. The effect of the titling policy on the propensity to invest and on the value of investments is positive and significant for both groups, but shows a stronger impact on parcels with previously weaker levels of tenure security. Moreover, this effect can be almost entirely attributed to changes in farmer's willingness to invest and not on better access of credit.

Even though farmers can get access to informal land documents to increase their security over the land, we show that this procedure is mostly limited to farmers that were already better-off and it constitutes at best an imperfect substitution to the acquisition of full-fledged property titles like the ones provided by the PETT program. Our results highlight the importance of a public intervention like this one to lift-up the limitations for certain farmers to acquire tenure security by informal means. The recognition of informal land rights and the reliance of the program on community networks before the formalization of rights also appear to be fundamental for a successful intervention.

6.2.3 Credit constraints in the agricultural sector

The provision and registration of land titles has been hypothesized to have a direct impact on farmer's access to credit because of its effect on increasing the collateral value of land for credit lenders. This effect would be especially true regarding formal credit sources which often have imperfect information on borrowers and thus insist on collateral before advancing a loan. However, a large amount of evidence suggest a weak or even null impact of titling programs on credit access, particularly in Latin America (Boucher *et al.*, 2004; Guirkinger and Boucher, 2006). In addition, in the few cases where a positive effect could indeed be established it was found to be mostly in favor of wealthier producers (Aldana and Fort, 2001; Carter and Olinto, 2003).

Chapter 4 provides relevant information to understand the weak performance of the formal credit market in rural areas of Peru, and in particular the failure of the land titling policy to improve this situation. Even though formal credit institutions seem to be increasing their operations in areas with higher levels of titling density, due to an increase in the potential demand for loans and the higher 'liquidity' of land markets, there are other important limitations for farmers to access these loans even in these areas.

More than half of our sample of farmers is non-price rationed in the credit market, with the highest percentage reporting a lack of sufficient amounts of land and collateral value as main reason for being rejected or self-excluded. The rest restrained themselves from participating in this market, even though they would have liked to apply for loans, either because they perceive the loan contracts as bearing too much risk for them, or they lack proper information and fear the high transaction cost embedded in the application process.

Even though land titling is supposed to increase the collateral value for land and thus reduces the probability of a farmer being quantity rationed, we do not find support for this effect. Having a registered title does, however, reduce the transaction cost involved in formal loan applications as a result of the administrative simplification process included in the program.

Finally, the existence of all these limitations implies that getting access to a formal loan becomes almost exclusively an option for wealthier farmers, with large amounts of land, and high levels of education. Land titling could facilitate their access and probably improve the conditions of their borrowing contracts, but it does hardly affect the possibilities for small-scale and poor producers.

6.2.4 The externality effect of titling on investments and land values

Land titling programs are usually based on the assumption that full-fledged land rights provide incentives and opportunities for individual farmers to invest in improved resource-use strategies. Current approaches devote little attention to the importance of a scale of titling and to the potential role of externalities for the development of local factor markets. Chapter 5, therefore, explores the implications of a new possible impact of titling on land investments and land values derived from an 'externality effect' that emerges with an increase on the number of titled plots in the same district (titling density).

Our evidence suggests that - although individual titling and registration can increase the level of farmer's tenure security and as such contribute to the enhancement of land related investments in their parcels - the strength of this relationship is very much related to the density of land rights formalization in the area where parcels are located. This factor is of particular importance for stimulating investments that have a high contribution in raising the value of land.

We showed that, apart from its effect on investments, individual title and registration and the level of titling density have a positive impact on farmer's perception about the price of their land. Individual titling and registration reinforces tenure security and helps reducing transaction and enforcement costs, while increased levels of titling density reflect expanded market opportunities for land.

These findings indicate, on the one hand, that for individual titling policies to become effective, other conditions to reduce transaction costs in rural areas and to improve the dynamics of land markets need to be fulfilled while, on the other hand, these conditions can at least be partially improved when the levels of titling density start to increase. However, this policy alone might not be sufficient for lifting up other important constraints in rural markets. The relative small change in the probability of investment, and on the investment value, reveals that there is still a positive demand for land-related investments (in particular for infrastructures) amongst many farmers that cannot be satisfied mostly because of a lack of financial opportunities. When asked about their limitations, around 70 percent of the farmers revealing a positive demand mentioned the lack of credit as their principal constraint to undertake investments. Therefore, even if formal financial institutions would be more willing to locate themselves and provide credits in Districts with higher levels of titling density, binding constraints from the demand side -- as the lack of information or insurance against negative shocks, and the extremely small size of land-holdings -- need to be addressed in order to enhance the potential benefits of this intervention.

6.3 Main Implications and policy recommendations

In the light of the findings described above, titling and registration policies can be considered as a necessary condition to improve investment opportunities when its implementation procedure is based on the recognition of previous informal land rights and community networks, and also because of its effect on the reduction of transaction costs at a regional level, which improves the dynamics of land markets and facilitates the entrance of formal financial institutions. A decentralized program is more likely to understand and correctly assess local conditions, as well as to concentrate its work on poorer farmers confronting stronger limitations to acquire tenure security by other means. Targeting must be applied also at the regional level, identifying less-developed areas that can benefit from the externality effects provided by increased levels of titling density.

However, the presence of other limitations that constrain the participation of small farmers in the formal credit market, and the failure of titling to solve them by itself, makes it difficult to consider this policy as a sufficient condition to improve the livelihood of poorer farmers. Moreover, the lack of long-term credit access in spite of titling could affect their competitiveness in the land market, reducing their chances to increase their scale of operations, and might even force them out of the agricultural sector.

Hence, complementary policies that provide small farmers with the opportunity to increase their land-holdings, as well as the possibility to acquire insurance against negative shocks, need to be urgently implemented. Promoting associative schemes amongst small producers might be an option to consider. However, given the negative experience with the mandatory associations created after the Agrarian Reform, any policy design would have to be carefully studied as to create the right incentives for farmers to group together.

The possibility for farmers to acquire some type of insurance contract that could help them to confront the usual negative shocks in agriculture deserves further analysis. Apart from reducing farmer's vulnerability, it might also contribute to increase their willingness and ability to undertake investments and increase their productivity. While conventional agricultural insurance schemes maintain adverse selection and moral-hazard limitations, new experiments on area-based yield insurance have to be analyzed. Other options that increase the diversification capacity of small producers could also contribute to improve their risk-bearing profile.

6.4 Future research

In order to explore in more detail the conclusions derived from this study, some issues require further research. In terms of the analysis at the country level, it would be desirable to expand the sample of countries with accurate information about changes in land distribution, particularly to include a larger number of underdeveloped countries, so that more instruments and controls can be used in the analysis. Another option would be to obtain a broader measure of assets distribution (i.e. including housing and urban land ownership). It could also be important to find measures that reflect more directly ownership security. Many other factors - such as social interaction problems, political instability or ethnic heterogeneity - can also be helpful to better understand the mechanisms by which assets inequality affects economic growth.

In terms of the effect of individual titling and titling density on reducing transaction and enforcement costs in the land market, more work needs to be done in order to outline the pattern of actual transactions and its implications in terms of efficiency and equity goals. An effort has to be made to collect more and better information at the meso-level and to explore its relationship with household's behavior within that region.

Finally, although this study had to sort out the traditional limitation of dealing only with a cross-sectional database, a new visit to the same sample of household's may provide a better source of information for testing more complex relationships and avoid potential methodological problems. It will be important to continue this initiative so that a longer panel data-set becomes available to allow testing for long-term effects of the Land Titling Program.

References

Aghion, P., C. Eve and C. Garcia-Penalosa (1999). Inequality and Economic Growth: The Perspective of the New Growth Theories. Journal-of-Economic-Literature. December 1999; 37(4): 1615-60(0022-0515).

Aguilar, G. (2004). El impacto del AGROBANCO sobre las microfinanzas rurales. SEPIA X: El probelma Agrario en Debate, Pucallpa.

Aldana, U. and R. Fort (2001). Efectos de la titulacion y registro sobre el grado de capitalizacion en la agricultura peruana. Economia y Sociedad. N.42. Lima, CIES.

Alesina, A. and D. Rodrik (1994). Distributive Politics and Economic Growth. Quarterly Journal of Economics. May 109(2): 465-90.

Arellano, M. and O. Bover (1995). Another Look at the Instrumental Variable Estimation of Error-Components Models. Journal of Econometrics. July 68(1): 29-51.

Arellano, M. and S. Bond (1991). Some test of specification for panel data: Monte Carlo evidence and an application to employment equations. The Review of Economic Studies. 58: 277-97.

Atwood, D. A. (1990). Land Registration in Africa: The Impact on Agricultural Production. World Development. May 18(5): 659-71.

Barham, B. L., S. Boucher and M. R. Carter (1996). Credit Constraints, Credit Unions, and Small-Scale Producers in Guatemala. World Development. May 24(5): 793-806.

Barro, R. J. and J.-W. Lee (2000). International Data on Educational Attainment Updates and Implications. NBER Working Papers 7911.

Barrows, R. and M. Roth (1989). Land Tenure and Investment in African Agriculture: Theory and Evidence. LTC Paper N.136. Madison, Land Tenure Center.

Barzel, Y. (1989). Economic analysis of property rights. Cambridge, Cambridge University Press.

Besley, T. (1995). Property Rights and Investment Incentives: Theory and Evidence from Ghana. Journal of Political Economy. October 103(5): 903-37.

Binswanger, H. P., K. Deininger and G. Feder (1995). Chapter 42: Power, Distortions, Revolt and Reform in Agricultural land Relations. Handbook of Development Economics. J. Behrman and T. N. Srinivasan, Elsevier Science. III.

Birdsall, N. and J. L. Londono (1998). No Tradeoff: Efficient Growth via More Equal Human Capital Accumulation. Beyond tradeoffs: Market reforms and equitable growth in Latin America. N. Birdsall, C. Graham and R. Sabot. Washington, D.C, Inter-American Development Bank: 111-45.

Birdsall, N., D. Ross and R. Sabot (1995). Inequality and Growth Reconsidered: Lessons from East Asia. World Bank Economic Review 9(3): 477-508.

Boucher, S. (2000). Information Asymmetries and Non-price Rationing: A Theoretical and Empirical Exploration of Rural Credit Markets in Northern Peru. Applied and Agricultural Economics. Madison, Wisconsin, University of Wisconsin-Madison. Ph.D.

Boucher, S. and M. R. Carter (2002). Risk rationing and activity choice in moral hazard constrained credit markets, University of Wisconsin.

Boucher, S., B. Barham and M. R. Carter (2004). The impacts of 'market-friendly' reforms on credit and land markets in Honduras and Nicaragua. World Development 33(1): 107-28.

Bourguignon, F. (1998). Crime, Violence, and Inequitable Development. Annual World Bank Conference on Development Economics, Washington, D.C. World Bank.

Bromley, D. W. (1998). Property regimes in economic development: lessons and policy implications. Agriculture and the Environment. E. Lutz. Washington D.C., The World Bank.

Bruce, J. W. (1986). Land Tenure Issues in Project Design and Strategies for Agricultural Development in Sub-Saharan Africa. Land Tenure Center. LTC Paper N.128

Bruce, J. W. and S. E. Migot-Adholla (1994). Serching for land tenure security in Africa. Iowa, Kendal/Hunt Publishing Cy.

Byamugisha, F. (1999). The effects of land registration on financial development and economic growth: a theoretical and conceptual framework. World Bank Policy Research Working Paper 2240. T. W. Bank.

Caballero, J. M. (1980). Agricultura, Reforma Agraria, y Pobreza Campesina. Lima, IEP.

Carter, M. R. (1988). Equilibrium credit rationing of small farm agriculture. Journal of Development Economics 28(1): 83-103.

Carter, M. R. (2000). Land Ownership Inequality and the Income Distribution Consequences of Economic Growth. Working Paper N.201. WIDER, The United Nations University.

Carter, M. R. and P. Olinto (2003). Getting Institutions 'Right' for Whom? Credit Constraints and the Impact of Property Rights on the Quantity and Composition of Investment. American Journal of Agricultural Economics. February 85(1): 173-86.

Carter, M., K. D. Wiebe and B. Blarel (1994). Tenure Security for whom? Differential effects of land policy in kenya. Serching for land security in Africa. J. W. Bruce and S. E. Migot-Adholla. Iowa, Kendal/Hunt Publishing.

Chung, I. (1995). Market choice and effective demand for credit: the roles of borrower transaction costs and rationing constraints. Journal of Economic Development 20(2): 23-44.

Collier, P. (1983). Malfunctioning of African Rural Factor Markets: Theory and a Kenyan Example. Oxford Bulletin of Economics and Statistics. May 45(2): 141-72.

Collier, P. (1998). The Political Economy of Ethnicity. Annual World Bank Conference on Development Economics, Washington, D.C.: World Bank,.

De Alessi, L. (1980). The Economics of Property Rights: A review of the evidence. Research in Law and Economics 2(1): 1-47.

De Janvry, A. (1981). The Agrarian Question and Reformism in Latin America. Baltimore, The Johns Hopkins University Press.

De Janvry, A. and E. Sadoulet (1995). Quantitative Development Policy Analysis. Baltimore, The Johns Hopkins University Press.

Deininger, K. and H. Binswanger (1999). The Evolution of the World Bank's Land Policy: Principles, Experience, and Future Challenges. World Bank Research Observer. August 14(2): 247-76.

Deininger, K. and J. S. Chamorro (2004). Investment and Equity Effects of Land Regularisation: The Case of Nicaragua. Agricultural Economics. March 30(2): 101-16.

Deininger, K. and L. Squire (1998). New Ways of Looking at Old Issues: Inequality and Growth. Journal of Development Economics 57(2): 257-85.

Deininger, K. and P. Olinto (1999). Asset distribution, inequality, and growth, World Bank.

Deininger, K., E. Zegarra and I. Lavadenz (2003). Determinants and Impacts of Rural Land Market Activity: Evidence from Nicaragua. World Development 31(8): 1385-1404.

Demsetz, H. (1967). Towards a Theory of Property Rights. American Economic Review 57(2): 347-359.

Dominicis de, L., H. d. Groot and R. Florax (2006). Growth and Inequality: A Meta-Analysis. Tinbergen Institute Discussion Paper N.2006-064

Easterly, W. and S. Rebelo (1993). Fiscal Policy and Economic Growth: An Empirical Investigation. Journal of Monetary Economics. December 32(3): 417-58.

Escobal, J. (1998). Nuevas Inversiones en el Agro de la Costa, GRADE.

Escobal, J. (2006). Desarrollando mercados rurales: El rol de la incertidumbre y la restriccion crediticia. Documento de Trabajo. GRADE. Lima, GRADE.

Feder, G. and D. Feeny (1991). Land Tenure and Property Rights: Theory and Implications for Development Policy. World Bank Economic Review. January 5(1): 135-53.

Feder, G., J. Y. Lau, Y. Lin and X. Luo (1990). The Relationship Between Credit and Productivity in the Chinese Agriculture: A microeconomic Model of Disequilibrium. American Journal of Agricultural Economics: 1151-1157.

Feder, G., T. Onchan, Y. Chalamwong and C. Hongladarom (1988). Land policies and farm productivity in Thailand. Baltimore, Johns Hopkins University Press.

Field, E. (2005). Property Rights and Investment in Urban Slums. Journal of the European Economic Association. April May 3(2-3): 279-90.

Forbes, K. J. (2000). A Reassessment of the Relationship Between Inequality and Growth. The American Economic Review 90(4): 869-887.

Galor, O., O. Moav and D. Vollrath (2004). Land inequality and the origin of divergence and overtaking in the growth process: theory and evidence, Brown University.

Greene, W. H. (2003). Econometric Analysis. NJ, Prentice Hall.

Guirkinger, C. (2005). Risk and the persistence of informal credit in rural Peru, Agricultural and Resources Economics. University of California Davis.

Guirkinger, C. and S. Boucher (2006). Credit constraints and productivity in Peruvian agriculture, Agricultural and Resources Economics. University of California Davis.

Hausman, J. and W. E. Taylor (1981). Panel Data and Unobservable Individual Effects. Econometrica 49(6): 1373-1399.

Keefer, P. and S. Knack (2002). Polarization, Politics and Property Rights: Links between Inequality and Growth. Public Choice. March 111(1-2): 127-54.

Knack, S. and P. Keefer (1995). Institutions and Economic Performance: Cross-Country Tests Using Alternative Institutional Measures. Economics and Politics. November 7(3): 207-27.

Kuznets, S. (1955). Economic Growth and Income Inequality. The American Economic Review 45(1): 1-28.

Larson, J. M., S. M. Smith, D. G. Abler and C. Trivelli (2003). An economic Analysis of land titling in Peru. Quarterly Journal of International Agriculture 42(1): 79-97.

Li, H. and H. f. Zou (1998). Income Inequality Is Not Harmful for Growth: Theory and Evidence. Review of Development Economics. October 2(3): 318-34.

Libecap, G. D. (1989). Distributional Issues in Contracting for Property Rights. Journal of Institutional and Theoretical Economics. March 145(1): 6-24.

Lindert, P. H. (1996). What Limits Social Spending? Explorations in Economic History. January 33(1): 1-34.

Lipton, M. (1974). Towards a Theory of land Reform. Peasants, Landlords, and Governments: Agrarian Reform in the Third World. D. Lehman. New York, Holmes & Meyer.

Lopez, R. (1996). Land Titles and Farm Productivity in Honduras, University of Maryland. Department of Agricultural and Resource Economics.

Madison, University of Wisconsin.

Matos Mar, J. (1980). Reforma Agraria: Logros y Contradicciones 1969-1979. Lima, IEP.

Meyer, B. D. (1995). Natural and Quasi-Experiments in Economics. Journal of Business & Economic Statistics 13(2): 151-161.

Migot Adholla, S. (1991). Indigenous Land Rights Systems in Sub-Saharan Africa: A Constraint on Productivity? World Bank Economic Review. January 5(1): 155-75.

Mushinski, D. (1996). Microenterprise and Small Business Access to Credit. AAE. Madison, University of Wisconsin-Madison. PhD.

Mushinski, D. (1999). An analysis of loan offer functions of banks and credit unions in Guatemala. Journal of Development Studies 36(2): 88-112.

Noronha, R. (1985). A review of the Literature on Land Tenure Systems in Sub-Saharan Africa. Report No: ARU 43. Washington, The World Bank.

Norton, E. C., H. Wang and C. Ai (2004). Computing interaction effects and standard errors in logit and probit models. The Stata Journal 4(2): 154-167.

Okoth-Ogendo, H. W. O. (1976). African Land Tenure Reform. Agricultural Development in Kenya: An Economic Assesment. J. Hayer, J. K. Maitha and W. M. Senga. Nairobi, Oxford University Press: 124-142.

Perotti, R. (1996). Growth, Income Distribution, and Democracy: What the Data Say. Journal of Economic Growth. June 1(2): 149-87.

Persson, T. and G. Tabellini (1994). Is Inequality Harmful for Growth? American Economic Review. June 84(3): 600-621.

Platteau, J. P. (1992). Land Reform and Structural Adjustment in Sub-Saharan Africa: Controversies and Guidelines. FAO Economic and Social Development Papers N.107. Rome, FAO.

Platteau, J. P. (1996). The Evolutionary Theory of Land Rights as Applied to Sub-Saharan Africa: A Critical Assessment. Development and Change. January 27(1): 29-86.

Platteau, J. P. (2000). Institutions, social norms, and economic development. Reading, U.K., Hardwood Academic.

Portocarrero, F. and A. Tarazona (2003). Determinantes de la Rentabilidad en las Cajas Rurales de Ahorro y Credito. Mercado y Gestion del Microcredito en el Peru. C. Trivelli. Lima, CIES.

Ray, D. (1998). Development Economics. New Jersey, Princeton University Press.

Rodrik, D. (2000). Institutions for High-Quality Growth: What They are and How to Acquire Them. NBER Working Papers(7540).

Rosen, S. (1974). Hedonic prices and implicit markets: product differentiation in pure competition. Journal of Political Economy 82(1): 34-55.

Roth, M. (1993). Somalia Land Policies and Tenure Impacts: The Case of the Lower Shebelle. Land in African Agrarian Systems. T. Bassett and D. Crummey. Madison, The University of Wisconsin Press: 298-325.

Stiglitz, J. E. (1969). The distribution of Income and Wealth among individuals. Econometrica 37(3): 382-97.

Stiglitz, J. E. and A. Weiss (1981). Credit Rationing in Markets with Imperfect Information. The American Economic Review 71(3): 393-410.

Temple, J. (1998). Initial conditions, social capital and growth in Africa. Journal of African Economies 7: 309-47.

Van Tassel, E. (2004). Credit Access and Transferable Land Rights. Oxford Economic Papers. January 56(1): 151-66

Wooldridge, J. (2002). Econometric Analysis of Cross Section and Panel Data. Cambridge, Massachusetts, The MIT Press.

World-Bank (1997). World Development Report 1997: The State in a Changing World. Washington D.C., World Bank.

Zegarra, E. (1999). El mercado de tierras rurales en el Peru. Serie Desarrollo Productivo. CEPAL. Santiago de Chile, CEPAL: 56.

Zimmerman, F. J. and M. R. Carter (1999). A Dynamic Option Value for Institutional Change: Marketable Property Rights in the Sahel. American Journal of Agricultural Economics. May 81(2): 467-78.

Zoomers, A. and G. v. d. Haar, Eds. (2000). Current Land Policy in Latin America. Regulating land tenure under neo-liberalism. Amsterdam, Royal Tropical Institute.

Summary

The legal framework regarding land issues in Peru as well as the economic policies for the agricultural sector have radically change during the last three decades, from a strongly regulated and interventionist approach, to a more liberal and free-market perspective. After the Agrarian Land Reform of the 1970's, and the parcelation process and subsidized schemes of the 1980's, the rural sector in Peru was confronted with market liberalization reforms in the 1990's, which took place in a context of extreme land fragmentation, low productivity levels, and an extended lack of formal and clear documents of ownership over the land.

Under the neo-classical paradigm, property rights reforms - which assign legally-secure and usually marketable land rights to farmers – as well as a constructive engagement with land markets, appear as fundamental instruments for contemporary land policy. Theoretical supporters of land titling and registration programs assert that well-established property rights and organized systems for public registration of property are an essential condition to improve the dynamics of land markets and to move towards a more efficient distribution of resources. At the same time, poor households will be able to use their secure assets as collateral for loans and will have a security-induced incentive to invest in its improvement, contributing in that way to increase the market value of their property and improve their competitiveness in the land market.

However, so far the empirical evidence supporting these arguments is largely inconclusive. The presence of multiple market imperfections in recently liberalized rural economies, and the subsistence of informal or customary property rights, seems to determine in practice whether or not these effects materialize, their relative importance, and also its consequences in terms of efficiency as well as equity goals.

This study discusses the links between land access, property rights, and economic development, analyzing the results and limitations of a public intervention- Land Titling and Registration- that constitutes one of the main instruments for contemporary land policy in Peru. It starts with a global perspective, and then proposes a meso (or regional) and micro level approach for the study of the Peruvian Land Tilting and Registration Program (PETT). The study attempts to provide a comprehensive analysis and discussion of the importance of institutions, like land property rights, in the context of market liberalization reforms. In operational terms, this means verifying whether land titling constitutes a necessary and/or sufficient condition to promote investments and increase land values. To accomplish this objective, we use information at two different levels. We assembled a country-level panel dataset for the macro perspective, and rely on household's surveys collected during the year 2004 as part of the evaluation of the PETT Program for the micro approach of this study. Our main research questions are the following:

1. What is the role of property rights in shaping the relationship between land distribution and economic growth?
2. How do legal documents affect farmer's tenure security and land-related investments? Is land titling required to enhance this effect?
3. What are the principal determinants and constraints that farmers face for accessing to formal sources of credit? Can land titling lift up some of these impediments and improve credit access for its beneficiaries?
4. Can land titling programs generate an externality effect on investments and land values by increasing the regional coverage of land rights formalization?

Chapter 2 provides a theoretical discussion and empirical evidence to better understand the relationships between land distribution and economic growth accounting for the role of secure property rights. Using an innovative panel data set with changes in land distribution over time and between countries, we have been able to provide confirmation for the hypothesis that asset distribution is a major determinant of economic growth. Apart from a direct effect, we also show that land inequality creates a barrier to the effectiveness of educational policies. Moreover, the incorporation of the physical investments variable in the model corroborates the existence of a growth-reducing impact of land inequality that goes beyond the conventional channel of credit market imperfections and reduced investments. Even though the security of property rights emerges as an important factor to explain economic growth, its effect does not modify the relationship found between land inequality and growth. These results have two important implications for policy strategies. First, it becomes clear that policies aiming at a more equal distribution of assets will be more effective if combined with complementary measures towards educational reforms and the improvement of institutional arrangements towards secure property rights. The lack of such a combined implementation of structural reforms can be one of the reasons why land reforms in several countries failed in the past to achieve the expected economic growth. Second, for developing countries that pursue market liberalization and privatization programs, it becomes of fundamental importance to remain alert that the effects of these reforms are not leading to the concentration of assets in few hands. Such unintended consequences are likely to deteriorate the country's economic performance in the long run.

In chapter 3, we use retrospective information regarding the type of informal documents that parcels had before the start of the PETT program to categorize them into two different levels of initial tenure security. The effect of titling on investments is then analyzed for these two groups of parcels using a difference-in-difference estimation technique. Our results show that there is a positive effect of titling on the probability of making investments, as well as on the value of investments for both groups of parcels, but also prove that its impact is higher for parcels with previously low levels of tenure security. Moreover, this effect can almost entirely be attributed to changes in farmer's willingness to invest and not to better access of credit. Even though farmers can get access to informal land documents to increase their security over the land, we show that this procedure is mostly limited to farmers that were already better-off and it constitutes at best an imperfect substitution to the acquisition of full-fledged property titles like the ones provided by the PETT program. Our results thus highlight the importance of a public intervention like this one to lift-up the limitations for certain farmers to acquire tenure security by informal means. The recognition of informal land rights and the reliance of the program on community networks before the formalization of rights appear to be fundamental for a successful intervention.

Chapter 4 explores the characteristics of supply and demand for formal loans in the Peruvian agricultural sector, and analyzes the principal determinants and constraints that farmers face for accessing this source of credit. Even though formal credit institutions seem to be increasing their operations in areas with higher levels of titling density (% of titled and registered parcels in a district), because of an increase in their potential demand for loans and the higher 'liquidity' of land markets, there are other important limitations for farmers to access these loans, even in these areas. We use survey questions specifically designed to identify rationing mechanisms for each individual, and a multinomial logit regression to determine the probability of being subject to each type of rationing. Our results show that more than half of our sample of farmers has a positive loan demand that is unsatisfied because of the presence of information asymmetries. While having a registered land title appears to decrease the transaction cost involve in formal loan applications, it is far from being a sufficient condition to get access to a loan. The existence of multiple limitations from the supply and demand side implies that getting access to a formal loan becomes almost exclusively an option for wealthier farmers, with large amounts of land, and high levels of education. Land titling could facilitate their access and probably improve the conditions of their borrowing contracts, but it does hardly affect the possibilities for small-scale and poor producers.

Chapter 5 discusses and explores yet another possible impact of titling on individual investments and land values, derived from an externality effect that appears with an increase of the number of titled plots in the same district. The evidence presented suggests that - although individual titling and registration can increase the level of farmer's tenure security and as such contributes to the enhancement of land-related investments in their parcels - the strength of this relationship is very much related to the density of land rights formalization in the area where parcels are located. This factor is of particular importance for stimulating investments that have a high contribution in raising the value of land. We also showed that, apart from its effect on investments, individual title and registration and the level of titling density have a positive impact on farmer's perception about the value of their land. Individual titling and registration reinforce tenure security and help reducing transaction and enforcement costs, while increased levels of titling density lead to expanded market opportunities for land.

These findings indicate, on the one hand, that for individual titling policies to become effective, other conditions need to be fulfilled to reduce transaction costs in rural areas and to improve the dynamics of land markets while, at the same time, these conditions can at least be partially improved when the levels of titling density start to increase. However, this policy alone might not be sufficient for lifting up other important constraints in rural markets. The relatively small change in the probability of investment, and in the investment value, reveals that there is still a positive demand for land-related investments (in particular for infrastructures) among many farmers that cannot be satisfied mostly because of a lack of financial opportunities.

The present study contributes to the existing literature in a number of ways. First, this study adds to the literature on the relationship between inequality and economic growth in two different aspects. It replaces the commonly used measure of income inequality by a dynamic one of assets inequality (land Gini) which better reflects the arguments of most theoretical models. In addition, it incorporates new arguments that link inequality in assets and property rights institutions with economic growth, contributing to the discussion on the potential effects of redistributive policies as well as complementary interventions to guarantee its correct functioning.

Second, this study analyses the effect of titling on tenure security and investments while taking into account previous levels of tenure security provided by the earlier acquisition of different types of informal land documents. Because titling might have a differentiated effect on investments - depending on the initial level of tenure security - and since the acquisition of 'stronger' documents might be limited for poorer farmers, previous studies that do not account for it are likely to overlook important consequences of this policy.

Finally, this study adds a new dimension to the analysis of land titling policies by introducing the notion of 'titling density' in our framework. Current approaches devote little attention to the importance of scale in titling and to the potential role of externalities for the development of local factor markets. The insights gained from such analysis may call for the introduction of a new regional perspective in the promotion of land titling programs and complementary policies to improve the livelihoods of the rural poor.

This study concludes with a number of lessons to consider for future implementation of Land Titling and Registration programs in the region, and outlines complementary policies that are required to guarantee the potential impact of land titling for small farmers. Titling and registration can be considered as a necessary condition to improve investment opportunities when its implementation procedure is based on the recognition of previous informal land rights and community networks, and also because of its effect on the reduction of transaction costs at a regional level, which improves the dynamics of land markets and facilitates the entrance of formal financial institutions. A decentralized program is more likely to understand, and correctly assess local conditions, as well as to concentrate its work on poorer farmers confronting stronger limitations to acquire tenure security by other means. Targeting must be applied also at the regional level, identifying less-developed areas that can benefit from the externality effects provided by increased levels of titling density.

However, the presence of other limitations that constrain the participation of small farmers in the formal credit market, and the failure of titling to solve them by itself, makes it difficult to consider this policy as a sufficient condition to improve the livelihood of poorer farmers. Moreover, the lack of long-term credit access could affect their competitiveness in the land market, reducing their chances to increase their scale of operations, and might even force them out of the agricultural sector. Hence, complementary and innovative policies that provide small farmers with the opportunity to increase their land-holdings, as well as the possibility to acquire insurance against negative shocks, need to be urgently implemented.

Printed in the United States
by Baker & Taylor Publisher Services